Carol Moleski

THE LONG SHADOW
OF TEMPERAMENT

JEROME KAGAN *&* NANCY SNIDMAN

THE BELKNAP PRESS OF
HARVARD UNIVERSITY PRESS
Cambridge, Massachusetts
London, England
2004

Library of Congress Cataloging-in-Publication Data

Kagan, Jerome.
The long shadow of temperament / Jerome Kagan and Nancy Snidman.
    p. cm.
Includes bibliographical references (p.   ) and index.
ISBN 0-674-01551-7 (alk. paper)
1. Temperament in children—Longitudinal studies.   2. Inhibition in
children—Longitudinal studies.   3. Temperament—Longitudinal studies.
4. Inhibition—Longitudinal studies.   5. Nature and nurture—Longitudinal
studies.   I. Snidman, Nancy C.   II. Title.

BF723.T53K345 2004
155.4'1826—dc22   2004040319

To the families whose generosity and loyalty
made this project possible

ACKNOWLEDGMENTS

We wish to thank Mark McManis and Sue Woodward for the seminal contributions they made to the assessment of 11-year-olds, which is the central focus of this book. We are grateful to Robin Schacht and Vali Kahn for help with these evaluations, and to Paula Mabee and Laura Gibson for manuscript preparation. We are especially indebted to Susan Wallace Boehmer, who once again has made an unruly manuscript comprehensible; and we thank Elizabeth Knoll, also at Harvard University Press, for editorial advice and sponsorship of this project. The assessment at 11 years was supported by a grant from the Bial Foundation, and the earlier work by grants from the John D. and Catherine T. MacArthur Foundation, the W. T. Grant Foundation, and the National Institute of Mental Health.

CONTENTS

We boil at different degrees.

—RALPH WALDO EMERSON

> Keep your eye clear
> As the bleb of the icicle
> Trust the feel of what nubbed treasures
> Your hands have known

—SEAMUS HEANEY

Because people are the most salient objects in our environment, the human mind cannot help but notice differences in physical features and behavior, especially social styles, accomplishments, character, and mood. Every culture has invented a theory to explain this variation. Although these theories differ in the fine details, they can be placed in one of two groups. One set of explanations awards most of the power to events in the outside world that impinge on the growing child. The other assigns more influence, but not all, to features that are inherent in the individual at birth—a function of one's astrological sign, a pregnant mother's fright, or an inherited biology. The concept of temperament is an essential component of this last factor.

Five thousand years ago when communities were small, most members of a village ate the same foods, lived in similar residences, and shared a common body of knowledge and values. It was easy to suppose that the reason why some infants were irritable and some calm must have something to do with internal forces within the child. However, when cities containing different classes and ethnic groups grew larger and more numerous, it was easier to conclude that psychological variation was created by distinctive ways of raising children and by other experiences in the home. This view was especially attractive in societies that held egalitarian values and aspired to equal dignity for all their citizens because child-rearing practices could, in theory, be improved with education.

As young students growing up in America, we, along with most psychologists of that era, preferred to believe that life experiences—in the family and the larger world—shaped the distinctive personalities that become so evident by school entrance. When Jerome Kagan was a graduate student almost fifty years ago, he was certain that a child's biology had little relevance in explaining why some adolescents were bold and some timid. He recalls, with a little embarrassment, arguing this point with his mother, who, out of intuition, was equally convinced that the explanation lay in mysterious elements within the child itself.

But decades of watching children in and outside the laboratory have persuaded us that each infant is born with a profile of temperamental biases, and some are not easily eliminated by life events. Initially, these biases display themselves in behaviors anyone can see. But by adolescence, a conscious "will" is able to control the face shown to the world, and it is impossible to detect these early temperaments from a person's daily actions.

We have spent the last twenty-five years probing two of the large number of temperaments that, we believe, will eventually be discovered. And we have learned that some features of these biases stubbornly resist extinction and continue to affect a person's private moods. This book presents the evidence for this claim. The infants whose temperament we identified as *high-reactive* were more likely to become shy and timid children; we called them *inhibited*. Those infants whose temperament we identified as *low-reactive* were more likely to become bold and sociable children; we called them *uninhibited*. The second-century physician Galen of Pergamon named these two temperamental types melancholic and sanguine; Jung called them introverts and extraverts. The fact that these categories have survived for two thousand years suggests that they may correspond to nature's plan and are not simply clever inventions.

Of course, different life histories create different personalities in children born with the same temperament. But one's temperament imposes a restraint on the possible outcomes. A low-reactive infant might become a trial lawyer, investment banker, navy pilot, or criminal, but it is unlikely that he will become a frightened recluse. Condensed water vapor can, depending on local conditions, form a white billowy cloud, a mackerel sky, or a dense ground fog, but it cannot become an asteroid.

Two caveats are important. First, no temperamental bias determines a particular personality type. Rather, each temperament creates an envelope of potential outcomes, with some more likely than others. Second, having a temperamental bias does not mean that adolescents or adults are not responsible for their behavior. People with healthy brains are able to control their actions, even though they may be less able to impose equivalent control over their feelings. The ancient Greeks understood this principle, for their four types referred to moods—choleric, melancholic, sanguine, or phlegmatic—and not to the behaviors that might flow from these moods, whether aggressive, nurturant, loving, sociable, or dishonest.

Chapter 1 provides an overview of what we have learned from studying children between the ages of 4 months and 11 years. Chapters 2 and 3 explain the intellectual framework and historical background of our work. Chapters 4 and 5 focus on the behavior and biology of our 11-year-olds, supplying technical details and statistical analyses that will allow other scientists to judge the validity of our claims. And finally, Chapter 6 offers some broader speculations invited by the evidence. In over a quarter-century of research, we have continued to learn new things about children and have often been surprised by our observations. We hope our readers can share vicariously in the deep pleasure of understanding we have had the good fortune to enjoy because of the continued loyalty of our many participating families.

1

Over the past 25 years, we have been intrigued by two distinct categories of children who react to unfamiliar people, animals, and situations in different ways. Some stare for a few moments at a stranger and then continue with their activities. A smaller group remains subdued for a longer time or retreats to a parent. Mothers and fathers usually notice these traits by the second or third birthday, when their son or daughter meets unfamiliar children at the playground or in child care, and some conclude that their child is by nature shy or sociable.

American and European scientists who have studied such children longitudinally have discovered that a proportion of shy or sociable preschoolers retain their early disposition into adolescence and, in some cases, into the adult years. For example, middle-class men growing up in Berkeley, California, who as children had been very shy, established their careers, chose wives, and became fathers later in life than their more sociable peers. Eighteen-year-olds in New Zealand who had been categorized by observers as shy when they were 3 years old described themselves as cautious when facing new challenges or dangerous situations (Caspi and Silva, 1995; Caspi, Elder, and Bem, 1988).

The contrast between a cautious and a bold character can be found in many novels, in Shakespeare's plays, and in E. L. Thorndike's 1907 textbook, *The Elements of Psychology*, published almost a century ago. Thorndike called the traits *reflective* and *impulsive*. We use the terms *inhibited* and *uninhibited to the unfamiliar* to capture this opposed pair of behavioral tendencies, and we believe that they are the product of temperamental biases that can be detected in early infancy.

The concept of temperament, as we use it, implies an inherited physiology that is preferentially linked to an envelope of emotions and behaviors (though the nature of that link is still poorly understood). Almost all 1-year-olds will freeze in place for a few seconds when an unfamiliar adult wearing a mask comes into the room; almost all 10-year-olds will flee if a large dog runs toward them; and almost all adults will feel apprehensive as they board a plane at midnight in an ice storm. But some infants—a minority—remain immobile with the adult for a long time; some 10-year-olds will cry as they flee from the dog; and some adults will continue to worry throughout the flight. We believe, on the basis of the research described in this book, that inherited temperamental biases, combined with life experiences, create this variation in behavioral reaction to unfamiliar events.

We wish to be clear at the outset that an infant's temperament is only an initial potentiality for developing a coherent set of psychological characteristics. No temperamental bias determines a particular cluster of adult traits. Life experiences, acting in potent and unpredictable ways, select one profile from the envelope of possibilities. A young boy born with a temperament that favors bold, sociable behavior may become the head of a corporation, a politician, a trial lawyer, or a test pilot if raised by nurturing parents who socialize perseverance, control of aggression, and academic achievement. A child born with exactly the same temperament is at some risk for a criminal career if his socialization fails to create appropriate restraints on behaviors that violate community norms.

Each adult profile was influenced by the child's temperament, but neither was predetermined by it.

A study of plants illustrates the power of the environment to produce different observable profiles from genetically identical forms. Each of seven genetically different variants of an herb were cut into three pieces, and one piece from each genetic type was grown at one of three altitudes: sea level, 1,400 meters, and in high mountains. The height to which each of the seven variants grew depended on both its genotype and the place where it was planted. For example, one variant was the tallest of the seven types if grown at sea level, the shortest if planted at 1,400 meters, and of intermediate height if planted in the high mountains. It was impossible to predict the final height of the plant from knowledge either of its genetic pedigree or the place where it grew; both facts were necessary (Lewontin, 1995).

Origins

Our studies of inhibited and uninhibited children have many roots. The deepest originate about 50 years ago in a study of Caucasian adults born between 1929 and 1939 who, from infancy through adolescence, had participated in a study at the Fels Research Institute on the campus of Antioch College in Yellow Springs, Ohio. In 1957 Howard Moss analyzed the staff's descriptions of childhood behavior, while Jerome Kagan interviewed and tested these young adults, who were now in their third decade. The two psychologists then examined both sets of evidence to see whether any childhood behaviors persisted through the early adult years.

The most important discovery was that a small group of children who usually avoided unfamiliar people, objects, and events during the first 3 years preserved some derivatives of that bias as young adults. They were introverted, cautious, and dependent for emotional support on family, friends, or spouses. By contrast, the bolder, sociable children were extroverts who chose competitive, entrepreneurial vocations. A singularly unexpected finding was that the adults who had been the timid children had high, minimally variable heart rates, suggesting higher sympathetic tone in the cardiovascular system. This evidence was the origin of the concepts of inhibited and uninhibited children (Kagan and Moss, 1962).

Fifteen years later, Kagan, Richard Kearsley, and Philip Zelazo were reflecting on their longitudinal observations of 53 Chinese-American and 63 Caucasian infants born in Boston who had been participating in a study on the effects of daycare. Some of the children had attended an experimental daycare center from 3 through 29 months of age; others had been reared only at home; but all infants were observed on five separate occasions. The Chinese-American infants, whether attending the daycare center or raised only at home, were more inhibited in the second year than the Caucasian children. The Chinese toddlers stayed closer to their mothers in unfamiliar places, cried more often when temporarily separated from their parent, and were unusually wary when playing with an unfamiliar peer (Kagan, Kearsley, and Zelazo, 1978). The results from the Ohio adults and the Boston children in daycare motivated more direct investigations of the temperamental properties of inhibited and uninhibited children.

The first investigation began in 1979 when Cynthia Garcia-Coll filmed 117 21-month-old Caucasian infants, both first- and later-born, as they encountered unfamiliar people, objects, and situations in a laboratory setting (Garcia-Coll, Kagan, and Reznick, 1984). Crying, withdrawal, and the absence of spontaneous behavior with the unfamiliar examiner, as well as prolonged hesitation in approaching unfamiliar objects or people, were regarded as signs of an inhibited response to unfamiliarity. The 33 children who were avoidant in a majority of unfamiliar contexts were categorized as inhibited. The 38 children who approached the unfamiliar people and objects were classified as uninhibited. The remaining 47 children who were inconsistent in their behavior were not observed again. When the inhibited and uninhibited children encountered the same set of unfamiliar events several weeks later, a majority retained the style they had displayed on the first evaluation.

Sixty percent of this sample was observed again on two occasions when the children were 4 years old. The formerly inhibited, compared with the uninhibited, children were more subdued while interacting with an unfamiliar peer, had higher task-induced heart rates, and glanced frequently at the examiner. They were also described as shy and fearful by their mothers. When the inhibited children were shown a series of pictures, each illustrating a pair of people, they looked longer at the passive rather than the active member of the pair. When some children in this sample were observed during their first day in kindergarten, those classified as inhibited at 21 months were solitary and quiet and stared often at other children, while the uninhibited children showed the opposite behavior.

In a related study, Nancy Snidman (1989) filmed a large group of 31-month-old children as they played with another unfamiliar child of the same sex in a large room with both mothers present. At the end of the play session, a woman with a plastic cover over her head and torso entered the room and after a brief period of silence invited the two children to approach. The children who remained close to their mother and were reluctant to approach the adult stranger, initiate play with the unfamiliar child, or speak spontaneously were classified as inhibited. The children who showed the opposite set of traits were called uninhibited.

Twenty-six children (about 15 percent of Snidman's sample) were shy with the unfamiliar child and timid with the adult stranger; 23 children (14 percent) were sociable and bold. These two groups were assessed again when the children were 4, 5, 7, and 13 years old. Children classified originally as inhibited retained an inhibited behavioral style, while children who had been classified as uninhibited preserved a bold, sociable profile. When a stringent criterion for classification was applied, about one-third of the children preserved their original temperamental bias, and only 3 of the 49 children changed from inhibited to uninhibited or from uninhibited to inhibited.

The inhibited and uninhibited children in the Garcia-Coll and Snidman samples were evaluated again when they were between 12 and 14 years old. Carl Schwartz, a child psychiatrist who interviewed these adolescents, discovered that those who had been inhibited were more likely to show signs of social anxiety (Schwartz, Snidman, and Kagan, 1999). During a subsequent testing session with an unfamiliar woman examiner, the formerly inhibited children failed to elaborate their answers or to ask the examiner questions, and they rarely smiled following success on various cognitive tasks. The formerly uninhibited children, by contrast, talked frequently and occasionally laughed after failing a difficult test item. Further, when asked to remember a long series of numbers and perform a motor coordination task as rapidly as possible, the inhibited children had either larger heart-rate accelerations or greater muscle tension. Finally, the inhibited youth were a little more likely to possess a tall, lean body build, narrow face, and light blue eyes.

The entire corpus of evidence convinced us of the significance of these two temperamental biases. In 1986 we designed an infant assessment to see whether we could detect in the behavior of 4-month-olds early signs of a future inhibited or uninhibited profile. We could not code avoidance of unfamiliar events because young infants are not capable of displaying the behaviors we call inhibited or uninhibited. Two psychological competences must develop before children will avoid or show caution in response to an unfamiliar person, object, or situation. First, the infant must relate her perception of the unfamiliar event to schematic representations of the familiar experiences encountered in the past. Second, the infant must be able to hold both representations in working memory long enough to compare the two. An inability to understand an unfamiliar event often results in crying or other actions reflecting uncertainty. The human brain at 3 to 4 months of age is not mature enough to perform these cognitive operations.

Reliable signs of timidity or avoidance in response to unfamiliar events first appear in human infants between 6 and 9 months of age, usually in the form of a fear reaction to strangers. An avoidant reaction to novelty appears in wolf cubs and kittens around 1 month, and at 3 months in infant monkeys, whose brains mature more slowly than wolf cubs' but faster than humans'. These facts suggest maturational constraints on this class of behavior. However, children overcome their timidity more quickly than monkeys. When a 4-month-old monkey confronts another unfamiliar monkey of the same age, the two wait a long time before one approaches the other. A pair of 12-month-old children who do not know each other may stare for a while, but within a minute or two most pairs begin to play. Humans conquer their initial restraint around the unfamiliar more quickly than monkeys.

The amygdala, a small almond-shaped organ lying beneath the convoluted cerebral cortex, contiguous to the hippocampus, influences a person's reaction to novelty. It is the only brain structure that detects change in both the outside environment and the body and, in addition, can instruct the body to flee, freeze, or fight. Every sensory modality sends information to one or more areas of the amygdala, and each area, in turn, sends projections to sites in the brain and body that mediate emotions and actions, including the cerebral cortex, brain stem, and autonomic nervous system. The amygdala can be likened to a central command post whose mission is first to detect and then, if necessary, to react to unfamiliar or unexpected events. Therein lies a clue to understanding the behavior of children when they encounter the unfamiliar.

The behavior of housecats affirms the significance of the amygdala in monitoring reaction to the unfamiliar. About 1 in 7 housecats behaves like an inhibited child. It fails to explore unfamiliar places, usually withdraws from unfamiliar objects, and does not attack rats. This avoidant profile, which first appears at about 30 days of age, becomes a stable trait by 2 months, when the kitten's amygdala gains control of the circuits that mediate avoidant behavior. Timid kittens also show a larger increase in neural activity in the amygdala than bold kittens when they hear sounds that resemble the threat howl of another cat (Adamec, 1991).

It is probably not a coincidence that the basolateral area of the amygdala, which receives sensory information from many modalities, expanded as primates evolved. This growth is particularly clear in gregarious monkey and ape species that engage in complex social interactions with conspecifics. Their social organization relies on the ability to detect discrepant facial, vocal, and postural cues that might signal domination, attack, or the moment when it is safe to approach another member of the troop for nurture, play, or mating.

Humans, like their primate relatives, are exquisitely sensitive to changes in facial expression, voice, and posture that signify anger, empathy, fear, seduction, delight, or disapproval from another person. If the variation in behavior to unfamiliarity is partly due to amygdalar excitability, it follows that coding the reactions which accompany arousal of the amygdala in infants might provide an index of a temperamental bias. Projections from the amygdala to motor centers enable infants to pump or thrash their limbs, arch their back, and cry when aroused by unfamiliarity. Thus, we hypothesized that infants born with a neurochemistry that rendered the amygdala unusually excitable would display vigorous motor activity and crying when presented with unfamiliar stimuli. Conversely, infants born with a higher threshold of reactivity in the amygdala due to a different chemistry would show minimal motor activity and distress in reaction to the same unfamiliar events.

The Infant Assessments

The central assumption behind our assessments of 4-month-old infants was that those who displayed frequent motor activity and crying when presented with unfamiliar visual, auditory, and olfactory stimuli had inherited a distinct neurochemistry that rendered the amygdala excitable. These infants should be biased to become inhibited children in the second year. The infants who showed minimal motor activity and little or no distress, because of a different neurochemistry, were likely to become uninhibited children.

This assumption was supported by LaGasse and colleagues (1989), who found that when the water newborn infants were sucking through a nipple suddenly turned sweet, some showed an immediate and large increase in rate of sucking. The unexpected change in taste sensation should have activated the amygdalae of all infants. Thus, those who showed a higher increase in sucking rate in response to the change in taste probably had a lower threshold of excitability in this structure. Two years later, these same children displayed shy and timid behavior in response to unfamiliar events. This finding implied an association between threshold of excitability in the amygdala and a temperamental bias to become inhibited.

In order to eliminate other conditions that might produce variation in infant behaviors, we recruited only infants born to mothers who had followed sound prenatal guidelines and had given birth at term to healthy infants without complications during pregnancy or delivery. In collaboration with Doreen Arcus (1991), we evaluated over 500 healthy Caucasian infants born to middle-class women, most of whom had college degrees. We selected 4 months as the time for the first assessment because by that age the amygdala and its projections are prepared to produce vigorous motor activity and distress when events fail to match the knowledge the infants had already acquired about the world.

FOUR-MONTH-OLD ASSESSMENTS

The 4-month evaluation, which lasted about 45 minutes, contained seven episodes that were discrepant from the infant's past experience. Initially, the mother looked down at her infant, smiling but not talking, for one minute. She then went to a chair behind the infant to be outside his field of vision. The examiner placed a speaker baffle to the right of the infant and turned on a tape recording that played 8 short sentences read by female voices. Most infants became quiet and remained alert during the presentation. However, some began to thrash their arms and legs, and a small number cried.

The speaker baffle was removed and the examiner, standing in back of the infant, presented a set of mobiles composed of 1, 3, or 7 unfamiliar colorful toys that moved back and forth in front of the infant's face for nine 20-second trials. Most infants were more active in response to the mobiles than to the taped sentences; some became increasingly aroused over the nine trials, as reflected in vigorous limb movements and crying. The examiner then dipped a cotton swab into very dilute butyl alcohol and presented it close to the infant's nostrils for 8 trials (the first and last trials were water rather than alcohol). The speaker baffle was replaced and the infant heard a female voice speaking three nonsense syllables *(ma, pa, ga)* at three different loudness levels. As before, most infants were quiet, but a small proportion thrashed and cried. The examiner then popped a balloon in back of the infant; most were unperturbed by this event. Finally, the mother returned to gaze at her infant for the final minute.

About 20 percent of infants showed crying and vigorous pumping of the legs and arms, sometimes with arching of the back, on at least 40 percent of the trials. These infants were called high-reactive. Because the motor activity and crying usually ceased when the stimulus was removed, it was reasonable to conclude that these reactions were caused by the unfamiliar stimulation rather than by hunger, transient pain, or sleepiness. The 40 percent who showed the opposite pattern—minimal motor activity and minimal distress—were called low-reactive. Occasionally these infants moved an arm or a leg, but they showed little crying or motor tension and appeared minimally aroused. About 25 percent showed low levels of motor activity but were very irritable during the stimulus presentations; these infants were called distressed. Finally, the smallest group, 10 percent, showed vigorous motor activity, usually pumping of arms and legs but no arching of the back, and rarely cried—they were called aroused. The remaining 5 percent of the sample were difficult to classify. We studied the developmental paths of these four groups over the next 10 to 12 years (see Kagan, 1994, for details of the initial assessments).

TWO-YEAR-OLD ASSESSMENTS

More than 300 infants from our original sample of 500 returned to our laboratory at 14 and 21 months, where they encountered a large number of unfamiliar events, including people, objects, procedures, and rooms. We coded the display of a fearful reaction to each unfamiliar stimulus. For example, we noted the child's resistance when the examiner placed heart-rate electrodes on her chest and a blood pressure cuff around her arm, and whether the child refused to put her finger into a small glass containing black fluid or refused to allow a drop of water to be placed on her tongue.

In another procedure, the examiner uncovered a rotating toy on the first two trials, smiled and uttered a nonsense phrase followed by the child's name (for example, *tat bubl, Marie*) in a friendly voice. On the second two trials she frowned while speaking the same words in a stern voice, and we noted whether the child cried following this change in the examiner's face and voice.

We coded the child's reaction to an unfamiliar woman dressed in a white laboratory coat wearing a gas mask, and we noted whether the child avoided a large metal robot that the examiner asked the child to approach. We also recorded whether the child was reluctant to approach a person dressed in a clown costume wearing a red and white mask, or a radio-controlled robot that moved while making sounds, despite invitations from the examiner to do so. Although all children recognized that these experiences were unfamiliar, only some showed the crying or avoidance that we regarded as behavioral signs of fear in response to the unfamiliar.

About one-third of the children showed either no fear or, at the most, one sign of fear during the varied episodes. Another third were very fearful, displaying a fear reaction to 4 or more situations. And one-third showed intermediate levels of fear. The children who had been high-reactive infants were most fearful, the low-reactives least fearful; the other two groups had intermediate fear scores. These findings convinced us that a temperamental bias contributed to the variation in reactivity at 4 months and the subsequent tendency to approach or avoid unfamiliar incentives in the second year.

FOUR-YEAR-OLD ASSESSMENTS

A sample of 193 of these children returned to the laboratory at 4.5 years for two sessions. During the first, which took about an hour, an unfamiliar woman administered a variety of cognitive tests. The children who had been low-reactive infants were more spontaneous and sociable with the examiner than those who had been high-reactive. Each child returned several weeks later for a play session in a larger room with two other unfamiliar children of the same sex and age while the three parents sat on a couch in the room. Twice as many low-reactive as high-reactive children were extremely sociable and talkative during this play session with unfamiliar children. By contrast, 46 percent of those who had been high-reactive infants were shy, quiet, and timid, compared with only 10 percent of those who had been low-reactive infants.

SEVEN-YEAR-OLD ASSESSMENTS

A group of 164 children were evaluated at 7.5 years for the presence of anxious symptoms. Initially, the mothers were sent a questionnaire which asked them to rate their child on a three-point scale for descriptions of age-appropriate behavior. Twelve questions dealt with shyness and timidity (for example, "my child becomes quiet and subdued in unfamiliar places"; "my child is afraid of thunder and lightning"; "my child is afraid of animals"). These answers were used to select a group of potential members of a category one might call anxious children. The mothers of the children selected as potentially anxious were interviewed on the telephone and asked to provide specific examples to support their earlier descriptions. These interviews revealed that some mothers had exaggerated the seriousness of their child's behavior. For example, when asked to explain why they said their child was afraid, the mothers indicated that the children were not exceptionally fearful. These children were eliminated from the category of potentially anxious children.

The teachers of children in the remaining group were interviewed. Each teacher, with no knowledge of the child's prior laboratory behavior or the purpose of the interview, first described and then ranked the child with respect to all children of the same sex in that classroom for characteristics that reflected shyness and fearfulness. The maternal questionnaire and telephone interviews with the mother and teacher were studied by Kagan, Snidman, and Marcel Zentner. If all three agreed that the child met the criteria for anxious symptoms, the child was categorized as anxious.

A total of 42 children (26 percent of the sample) met these criteria. A group of 107 control children were selected who did not meet criteria for any psychological symptom. An additional 15 children had symptoms of hyperactivity or disobedience but no signs of anxiety. All 164 children came to the laboratory for a battery of procedures administered by an unfamiliar woman who did not know their prior behavior.

The 7-year-olds who had been high-reactive infants were most likely to have anxious symptoms. Forty-five percent of high-reactives but only 15 percent of low-reactives received this classification. The high-reactives who developed anxious features were the ones most likely to have screamed in fear during the 21-month-assessment when a person dressed in a clown costume entered a room where they were playing. As at earlier ages, the high-reactives displayed fewer spontaneous comments and smiles while interacting with the examiner (Kagan and Snidman, 1999).

A total of 18 percent of the high-reactives were consistently inhibited at all four ages—14 and 21 months and 4.5 and 7.5 years—but not one high-reactive infant was consistently uninhibited across all four evaluations. Moreover, the high-reactives with anxious symptoms, compared with the non-anxious high-reactives, showed higher diastolic blood pressure and greater cooling of the temperature of the fingertips as they listened to a long series of numbers they had to remember. These latter two measurements imply a more reactive sympathetic nervous system in the presence of a cognitive challenge. And about one-fourth of the anxious high-reactives had a smaller body size and narrower face than the other children, as well as blue eyes.

ELEVEN-YEAR-OLD ASSESSMENTS

The results of the last assessment, which occurred when the children were between 10 and 12 years old, are the focus of the remaining chapters of this book. The sample of 237 children who returned to our laboratory contained 30 percent who had been high-reactive as infants and 39 percent who had been low-reactive; the remaining 31 percent belonged to the other two temperamental groups, originally called distressed or aroused. The laboratory evaluation, which took about 3.5 hours, measured a variety of biological variables, such as EEG power and asymmetry of activation, brain stem auditory-evoked potential, event-related potentials, and sympathetic reactivity in the cardiovascular system (changes in finger temperature, heart rate, and blood pressure and a spectral analysis of supine heart rate). These measures, described in detail in Chapter 4, were selected because each is potentially under the direct, or indirect, influence of the amygdala.

The evidence illuminates two questions:

- To what degree did the 11-year-old children from the temperamental groups display the behavioral and/or biological biases expected from their infant assessments?
- To what degree was their behavior (both past and present) yoked to their current biology?

The Behavioral Profile

We constructed different indexes of each preadolescent's behavior. One was based on social interaction with the examiner. We knew from past work that high-reactive children talked and smiled less frequently than low-reactives in unfamiliar social situations. All children should have been maximally uncertain during the initial part of the interview because both the examiner and the setting were new. Thus, we counted the number of times the child made a spontaneous comment or smiled at the examiner (who knew nothing of the child's past behavior) during the first 18 minutes of her interaction. A spontaneous comment was defined as an elaboration of an answer required by the examiner's question or an unprovoked comment or question made by the child. (We restricted the coding of comments and smiles to the first 18 minutes because the biological measures, gathered after that time, required the child to be still and quiet.)

A second behavioral index of inhibited and uninhibited behavior, more inferential, was an observer's judgment, on a 4-point scale, of the child's degree of inhibition over the first 18 minutes. The rating was based on infrequent comments and smiles, a soft voice, excessive muscle tension, nervous motor movements, gazing away from the examiner, and questions that reflected concern or worry over the procedures. A rating of 1 described a maximally uninhibited child who was relaxed, sociable, and spontaneous. A rating of 2 described a child who, although he did not smile and talk frequently, showed no other signs of uncertainty. A rating of 3 described a child who talked and smiled but showed other signs of uncertainty. A rating of 4 described a maximally inhibited child who rarely smiled or spoke, was tense, and displayed other signs of uncertainty.

A third source of evidence came from independent descriptions of the child provided by the mother and the child in the familiar home setting. Each was given different sets of cards with printed statements descriptive of the child. The mother was given 28 descriptions; the child was given 20. The mother and child, sitting in different rooms, arranged these descriptions from most to least characteristic of the child.

As anticipated, more children who had been high-reactive infants were quiet and serious while interacting with the examiner; more low-reactives were talkative and relaxed and smiled frequently. About 33 percent of the high- and low-reactives displayed a style of social behavior that was in accord with their infant temperament, while 16 percent of each group behaved in ways that were inconsistent with expectations—a ratio of 2 to 1. The number of spontaneous smiles was a particularly sensitive sign of infant temperament. More high-reactives preserved a serious, non-smiling facial expression at every assessment, from 14 months to 11 years. More low-reactives smiled and laughed frequently at every age. Many low-reactives, but very few high-reactives, smiled and laughed within the first minute of entering the laboratory at 11 years of age.

The mother's description of her child's usual behavior with strangers bore a modest relation both to infant temperament and to behavior with the examiner. About one-third of high-reactives were described by their mother as very shy around adult and child strangers; one-third of low-reactives were described as very sociable. However, the children's descriptions of their own shyness or sociability bore little relation to their early temperament, their behavior with the examiner, or their mother's evaluation. Many high-reactives who were quiet with the examiner and were described by their mothers as shy denied this quality in their self-descriptions. Many low-reactives who were sociable with the examiner and were described by their mothers as sociable reported being shy.

However, the children's descriptions of their usual moods and preference for novelty were linked to their infant temperaments. The low-reactives were most likely to report that they enjoyed new places to visit and novel experiences. One item in the child's self-descriptions was especially revealing. The low-reactives were more likely than high-reactives to report that they were "happy most of the time." We will see that most of these "happy" children had been low-reactive infants and had a distinct biology at 11 years of age.

To our surprise, the level of fear in response to unfamiliar events during the second year did not predict the child's behavior at age 11. That is, children who were very fearful at 14 and 21 months, across all temperaments, were not very different at age 11 from those who were minimally fearful. The infant's behavior at 4 months was a better predictor of their behavior and biology at age 11 than variation in fear at 1 and 2 years of age.

Smiling was the single exception to this generalization. Frequency of smiles in the second year did predict 11-year-old behavior. This fact suggests that moment-to-moment changes in emotion are more reliable indexes of high and low reactivity than avoiding or approaching unfamiliar events in the second year.

Yoking Current Biology to Behavior

Our selection of biological measures was guided by the hypothesis that high-reactives and low-reactives differ in amygdalar excitability. We chose a varied set of measures of cortical, brain stem, and autonomic reactivity that other research had implicated as potential signs of this property (see Chapter 4). Most of our expectations were confirmed. The biological measures that we regard as indirect signs of amygdalar excitability were more frequent among the 11-year-olds who had been high-reactive infants than among those who had been low-reactive. For example, the former showed greater activation of the right rather than the left hemisphere and a larger increase in cortical arousal in the EEG when the examiner asked them to reflect on their thoughts as they drove to the laboratory and then to stand up and recite them. They also displayed a larger evoked potential from the inferior colliculus (a target of projections from the amygdala), greater sympathetic tone in the cardiovascular system, and a larger negative waveform in the event-related potential in response to discrepant scenes.

The earlier observation that high-reactive children were more likely to have a thin body build, narrow face, and blue eyes was also affirmed in the 11-year-olds. A slightly shorter stature, lighter weight, and blue eyes were more common among the high-reactives, while more low-reactives were taller, heavier, and more often brown-eyed. The smaller body size of the high-reactive group could not be attributed to malnutrition or chronic disease because all children were well-fed and healthy. One possible explanation is that the genes contributing to infant reactivity, body size, and eye color are contiguous on a chromosome and travel together from one generation to the next during meiosis (the process which produces egg and sperm cells). A second, more likely possibility is that the genes that contribute to high or low reactivity are pleiotropic— that is, they influence many traits, some of which include body size, eye color, and timid or bold behavior in the face of unfamiliarity.

A study of silver foxes illustrates pleiotropy at work. A group of tame silver foxes, representing less than 10 percent of all the animals housed on a Siberian fox farm, were bred with one another—tame males were bred only with tame females. As a result of this selective breeding, changes in body form and fur pigmentation occurred as the mated animals produced increasingly large numbers of tame offspring. The offspring of the twentieth generation of the selective matings were not only tame but also displayed depigmented white spots in their fur and more flexible ears and tails, suggesting that the genes that contribute to tameness were pleiotropic. These changes in pigmentation and flexibility of ears and tail are characteristic of domesticated mammals, including goats, horses, and cows (Trut, 1999).

The modest relation between eye color and temperament in our young subjects is not an original discovery. Other psychologists have reported that shy school-age Caucasian children are more likely to be blue-eyed rather than brown-eyed. When teachers in 133 different classrooms nominated the Caucasian child in their class who was most shy and the Caucasian child who was least shy, 60 percent of the most shy children were blue-eyed and 58 percent of the least shy children were brown-eyed. A collaborative study with psychiatrists at the Massachusetts General Hospital revealed that among 148 Caucasian middle-class children with a parent who had panic disorder, panic disorder with depression, or no psychiatric condition, 10 girls born to a parent with panic disorder, usually the mother, were extremely shy with the examiner and had a very high heart rate—and all 10 girls had very light blue eyes.

An association between eye color and temperament lies quietly in the unconscious of the artists who have drawn figures for Walt Disney's animated films. Doreen Arcus (1989) discovered that the eye colors first given to characters portrayed as psychologically vulnerable—for example, Alice, Cinderella, and Dopey (one of the dwarfs in *Snow White*)—were blue. By contrast, the characters portrayed as aggressive or dominating—for example, the evil queen in *Snow White*, Grumpy (one of the dwarfs), the Queen of Hearts, Captain Hook, and Merlin—were drawn with dark eyes.

A number of implications flow from this rich set of evidence. The most interesting, and perhaps most significant, is that a 45-minute observation of the behavior of 4-month-old infants reveals a subset of children whose behavioral and biological characteristics could be anticipated to some degree over a decade later. This prediction was possible, we believe, because high- and low-reactive infants inherit different neurochemistries in the amygdala and its projections.

One-fourth of the preadolescents categorized at 4 months as high-reactives developed an expected pattern of behavior and biology. These 11-year-olds were described as shy by their parents, were quiet and subdued with the examiner, and displayed several of the biological signs that imply an excitable amygdala. And one-fourth of low-reactives preserved their expected behavioral and biological profile. But because 11-year-olds have considerable control over their public behavior, only a modest proportion of high-reactive preadolescents behaved in ways an observer would characterize as extremely shy, fearful, or timid, even though some of these non-shy youth showed physiological properties indicative of higher amygdalar activity.

In other words, the biology of the high-reactives had not prevented them from learning ways to cope with strangers and new challenges, but it did prevent them from displaying the relaxed spontaneity and low level of cortical and autonomic arousal characteristic of many low-reactive children. The fact that more than twice as many high-reactives as low-reactives had values on some of the physiological variables that imply an excitable amygdala, whether these adolescents were shy or bold, implies preservation of a biological property over the course of childhood.

The more important fact is that very few high-reactives became exuberant, sociable, minimally aroused preadolescents, and very few low-reactives became fearful, quiet introverts with high levels of biological arousal. Less than 5 percent of the children from each of these temperamental groups developed the behavioral and biological features characteristic of the other type. Most children who did not conform to expectation were neither extremely shy nor extremely exuberant, and few possessed more than one sign of amygdalar excitability.

The power of each infant temperamental bias lay in its ability to prevent the development of a contrasting profile. Put differently, the probability that a high-reactive infant would not become an ebullient, sociable, fearless child, with the appropriate biological profile, is very high. But the probability that a high-reactive infant would be extremely shy and show biological arousal at age 11 was much lower. The complementary pair of claims held for the low-reactives. An infant's temperament, therefore, constrains the acquisition of certain profiles more effectively than it determines the development of a particular personality.

Because variation in fear of the unfamiliar during the second year was less predictive of the 11-year profiles than was infant reactivity at 4 months, it appears that life experience had already begun to build a retaining wall around the young child's biology, constraining its influence on behavior. The environments in which children develop can create different personas in those who began life with the same temperamental bias. This process resembles the way that variation in climate, food supply, and predators can create, over time, different species from an original ancestor population. The fact that behavior at age 2 had already become a less transparent window into the child's temperament frustrates observers who would like to be able to predict the early temperaments of adolescents and adults from their current behavior.

Environmental Influences on Temperament

Many environmental conditions affect the future development of high- or low-reactive infants. Family socialization styles and values operate first. Parents who encourage boldness and sociability and gently discourage timidity motivate their high-reactive children toward a less inhibited profile. Doreen Arcus discovered that the mothers of high-reactive infants who showed minimal fear in the second year did not protect their infants from all uncertainty because they believed that their child had to learn to deal with unfamiliar challenges. The mothers of high-reactives who, out of equally loving concern, were reluctant to cause them distress and protected them from new experiences had the most fearful 2-year-olds.

Excessive parental criticism is also formative. Regular parental criticism can generate high levels of uncertainty in high-reactives, who are more vulnerable than low-reactives to feelings of anxiety or guilt following violation of family standards. Parental behaviors as role models are also influential. Most children identify with their parents; that is, they believe that some of the adult properties belong to them, and they experience vicarious emotion. Hence, a high-reactive girl with a bold, sociable parent is tempted to believe that she too has an outgoing persona. The high-reactive girl with a highly anxious mother comes to the opposite conclusion and might assume that her timidity is an inherent trait.

Actual success and failure, with people or with tasks, modulate development. A child with several satisfying friendships will have less doubt over her sociability than one with no close friends. Success in school, athletics, music, art, or other activities can persuade a highly reactive child that he has some basis for resisting a private judgment of inadequacy.

The specific group the child selects to compare himself to is critical for development. Shy children in contemporary American society would feel less uncertainty if they did not encounter so many sociable children. The mind/brain is exquisitely sensitive to infrequent events. Because extreme shyness is less common than sociability in the United States (but not necessarily in all societies), children and adults exaggerate the significance of this trait, and some view it as undesirable. The automatic attention paid to infrequent events is one reason why questionnaire measures of temperament are less valid than extensive behavioral observations. Parents of a shy child will attribute less shyness to that child if he possesses another even less common trait—for example, extreme aggression or impulsivity. By contrast, parents of a child for whom shyness is the only salient quality are tempted to exaggerate the seriousness of this trait.

The influences of temperament, past history, and current context are not additive. Each comes online at different times, and it is impossible to assign a weight to any one of these factors that reflects its unique contribution to the adolescent character. The temperamental bias is the first to exert its force, but the social environment begins to sculpt infants with varied temperaments into different profiles during the first year. And each child adjusts her psychological profile as settings of action change from home to playground to school to college to workplace. That is why it is difficult, at least at the present time, to discern an adult's early temperament by examining her current behavior or biology.

Unpredictable or chance events, too, have potency to change a life, or the lives of a group, just as a hurricane can eliminate a perfectly well-adapted population of mussels while enhancing the prospects of a competitive variety. Being first or later born in a family, growing up in a small town or big city, having competent or indifferent teachers at a critical age, or losing a parent in early childhood can alter the life course. Some unpredictable events, like war, civil unrest, or economic depression, can affect the ideas of millions of adolescents in a society and, in so doing, impact an entire generation. Two obvious examples for American youth were the Depression between 1930 and 1940 and the protest against the Vietnam War and racism in the late 1960s.

Yet, despite these facts, some temperamental biases can persist. Hox genes, which determine an animal's body architecture, offer an illuminating analogy. This small group of genes, which contributes to the bilateral body symmetry of worms, mosquitoes, frogs, lizards, hawks, mice, and humans, has been preserved for over 500 million years, while the genes for body size, type of skin covering, and shape of the limbs have been altered more drastically by natural selection working in concert with mutations and recombinations. We believe that the physiologies that are the foundation of high and low reactivity are more likely to be preserved over the life span than cognitive talents, political beliefs, or coping strategies. A person watching silent films of the first 20 minutes of interviews with these adolescents, and who understood the facts we have summarized, would, we believe, do better than chance in deciding which ones were high- and which ones low-reactive. But the viewer would be unable to predict that child's school grades, popularity with friends, emotional relation to parents, or athletic talents. The tell-tale signs would be the degree of muscle tension in the trunk and limbs and the frequency of smiling and laughter.

Advice for Parents

Most parents worry over the development of their children and implement practices they hope will protect their sons and daughters against undesirable outcomes. Educated parents in Westernized cultures are prone to believe that scientific facts should guide their rearing behavior. Although some suggestions of psychologists, pediatricians, and psychiatrists are no more valid than the superstitions of the ancient Greeks, we believe that some potentially useful advice flows from the evidence gathered on the high- and low-reactive temperaments.

First, parents should appreciate that some children inherit a temperamental bias to be cautious in unfamiliar situations. Parents should not assume that their parenting behavior is the sole source of this trait in their child's personality. An acceptance of that fact should alleviate some of the guilt felt by parents who misinterpret their child's timidity as a sign that they must have done something wrong. Conversely, proud parents of an ebullient and confident child should not assume full credit for their wise parenting skills. This suggestion does not mean that parents do not affect their child's personality—they do.

Parents should not overprotect high-reactive infants from all novelty or challenge, even though some loving parents are tempted to do so. Parents who gradually expose their high-reactive infants to feared targets can help them overcome their initial tendency to avoid the unfamiliar. Many of the inhibited children in our study gained control over outward signs of their uncertainty. One high-reactive inhibited boy wrote an essay for his teacher describing his eventual success in dealing with his temperamental bias.

I've always been more of an anxious person than some other people. There is a tendency for some people to be more shy and most of all more anxious. It took me a very long time to learn how to cope with this heightened state of nervousness. Until recently, I've been very scared of shots. One dentist refused to work with me. I finally found a dentist who I really trusted and then my fears ebbed away. I was also afraid to swim for a very long time because I was afraid of getting water on my face. I could not actually jump in until the day my sister did. When she jumped in the pool, I couldn't stand the feeling of inferiority. That made me jump in. From that day on, I could go under water. I used to have to talk to my parents all the time to calm down after a bad dream. However, I got over all those things. I sleep fine in the dark and have no problems with bad dreams.

I was able to get over my fears. These inner struggles pulled at me for years until I was able to just let go and calm myself. I have also found that the manifestation of my anxiety can be overcome by using simple mind-over-matter techniques. A good example of this was when I was 8 after learning about asthma, I started to feel like I was having trouble breathing. In a heightened state of anxiety, I subconsciously forced myself into believing that I had asthma. Besides just general fears, it was a struggle to overcome this anxiety manifestation. I overcame these problems. I know how to deal with them when they occur. For example, when I first heard about the anthrax in Washington, I began to have an upset stomach. I realized it was simply because of my anxiety that I was feeling sick. As soon as I realized that, the stomachache went away. Because I now understand my predisposition to-wards anxiety, I can talk myself out of simple fears.

PARENTING A HIGH-REACTIVE CHILD

Consider, as a thought experiment, three types of American fami-lies who bring a highly aroused, easily distressed infant home from the hospital. One class of parents feels empathic for this appar-ently unhappy infant and becomes overly solicitous. Mothers are likely to pick their infants up as soon as they cry and try to soothe the distress. This routine strengthens an infant's predisposition to cry at the first sign of novelty and, later, to avoid unfamiliar events. A reluctance to make their children unhappy motivates nurturing parents to accept a retreat from novelty, even though the accep-tance increases the likelihood that, as they become older, their chil-dren will be cautious in new situations. Continued acceptance of inhibited behavior can lead, in time, to shy, restrained, timid 6-year-olds. We estimate that about one-third of high-reactive Amer-ican infants experience this form of child-rearing.

A second group of equally loving parents holds a different phi-losophy. These families believe, often tacitly, that they must pre-pare their children for a competitive society in which retreat from challenge is maladaptive. These parents refrain from comforting their infants every time they cry and wait for the children to regu-late their own distress. Rather than accept retreat from unfamiliar-ity, these parents encourage their children to greet adult strangers and approach unfamiliar children on a playground, and they praise their sons and daughters when they overcome their caution. Children reared this way are less likely to be avoidant when it is time to begin first grade. Such children often display a high energy level, talk too much, and ask too many questions. This script is more common in sons than in daughters because of a sex-role bias in our culture that regards retreat from challenge as less acceptable in boys than girls. We suspect that this group, too, comprise about one-third of high-reactive infants.

A much smaller proportion of high-reactives are born to parents who misinterpret their irritability as willful. These parents often become angry with their children and punish what they regard as misbehavior or disobedience. This regimen exacerbates an already high level of limbic excitability. Because these children cannot always control their emotions, they can become more irritable or, depending on the severity of the parental behavior, withdraw as experimental animals do when they cannot avoid electric shock. Fortunately, this developmental course is far less common than the first two.

PARENTING A LOW-REACTIVE CHILD

Families with a low-reactive infant also vary in parenting styles. Low-reactive infants smile frequently, sleep well, and are easy to care for most of the time. However, when the first signs of self-awareness emerge during the second year, these children may resist parental demands for control of mild aggression, tantrums, disobedience, and destruction of property because parental chastisement does not generate as much uncertainty as it does in most 2-year-olds. At this point two developmental itineraries become possible.

In the most common, derived from the value Americans place on autonomy and freedom from anxiety, parents adopt a laissez-faire approach and permit disobedience to all but the most serious infractions. These children are likely to become relatively exuberant and sociable as long as their parents are consistent in socializing serious violations of community norms on aggression, domination, and destruction. However, if parents are inconsistent in their punishment and if, during later childhood, there are peer temptations for asocial acts, these children—especially boys—are at a higher than normal risk for developing an asocial personality.

A different course is likely if parents are unwilling to brook opposition to their demands and punish most disobedience. Children raised this way are likely to react with angry tantrums, and continuation of this cycle can create a rebellious posture toward parents and later toward the authority of teachers and employers.

The mothers' descriptions of their 11-year-olds affirm these hypothetical predictions. Fifteen percent of high-reactives, but only one low-reactive, were described as very shy with strangers; 30 percent of high-reactive boys, but only 15 percent of low-reactive boys, were described as having high energy and talking all the time. By contrast, 30 percent of low-reactives were described as exuberant, compared with only 5 percent of high-reactives, and 10 percent of low-reactive boys, but not one high-reactive boy, was described as rebellious.

A BALANCED PERSPECTIVE

Although many parents may assume that low-reactives represent the more desirable type of infant, each temperamental type has both advantages and disadvantages in our society. A technological economy requires a college education. Students with higher grade-point averages in high school tend to be accepted at better colleges and have a higher probability of attaining a gratifying, economically productive career. In middle-class homes, high-reactive children are more concerned with academic failure and therefore more likely to have an academic record that will gain admission to a good college. After college, this type of person will have many occupational options, because our society needs adults who like to work in environments where they can titer the level of uncertainty. Adolescents who were high-reactive infants often choose intellectual vocations because this work allows some control over each day's events in settings where unanticipated interactions with strangers can be held to a minimum. Our society needs these vocational roles, and those who fill them are rewarded with respect and financial security. I suspect that T. S. Eliot would have replied "no" if, after attaining international fame, he was asked whether he would have wanted his mother to take him to a therapist to treat his inhibited personality. Had Eliot conquered his extreme introversion, he might not have become a poet and playwright.

Contemporary adolescents are confronted with many temptations that promise pleasure, peer acceptance, and self-enhancement if they are willing to assume some risk. Driving at high speeds, experimenting with drugs, engaging in sex at an early age or unprotected sex at any age, and cheating on examinations are four common temptations with undesirable consequences; some decisions can change a life permanently. The adolescent who experiences uncertainty as he thinks about engaging in these behaviors—a more likely response among high-reactives—will tend to avoid the risk. It is not a coincidence that introverts live longer than extraverts. Guppies with a bold temperament approach large predator fish, while timid guppies stay away. The former have a shorter life span than the latter.

The low-reactive, uninhibited child enjoys her share of advantages. Sociability and a willingness to take career and economic risks are adaptive in contemporary American society. The 17-year-old willing to leave home in order to attend a better school or accept a more interesting job is likely to gain a more challenging position and greater economic return than one who stays close to home because of a reluctance to confront the uncertainties of a distant place. Had the parents of Lyndon Johnson and Bill Clinton tried to mute their sons' extreme extraversion, those men might not have become President.

The island of Cayo Santiago, close to mainland Puerto Rico, contains over 1,000 rhesus macaques and no human residents. Observers visit the small island each day to make notes on the animals' behavior. The consistently timid monkeys are likely to die of starvation because, when food is put out each day, they wait for the other monkeys to feed first. If a timid monkey waits too long on too many days, he will starve to death. On the other hand, the bold monkeys get sufficient food, but they are at a higher risk of dying from wounds that follow an impulsive attack on a stronger animal.

When scientists eventually discover drugs that can reduce the uncertainty of the inhibited child or rein in the fearlessness of the uninhibited, parents will have to decide whether to use these pills. Their decision will depend on the values of our society. If American culture 50 years hence is not very different from its present form, most parents of inhibited children will probably use the drug, while parents of uninhibited children will not. Should historical events place the uninhibited child at higher risk for adaptive failure, parents will make the opposite decision.

We end this overview with three suggestions for parents:

1. Acknowledge your child's temperamental bias and do not assume that either your rearing practices or the child's willfulness is the only reason for his or her behavior.
2. Acknowledge your child's malleability and capacity for change. An infant's biology does not determine what she will become 10 or 20 years later. Temperament is not destiny.
3. Accommodate parenting goals to the child's own wishes. A regimen of rearing that takes account of both the parent's hopes and the child's desires can be found, if parents are willing to search for it.

2

THE TAPESTRIES OF
TEMPERAMENT

Regularity and variation, like the complementary relation between melody and rhythm, are inherent in all of nature's compositions. The sun rises each morning but not in the same place or at the same time across the seasons. Each species begets its own kind, but a litter of puppies varies in coloration, size, and behavior. All cultures create ceremonies to mark the birth and death of their members, but the rituals are not identical. The causes of the regularities, whether physical, biological, or social, are different from the causes of the variation.

Similarly, the conditions that enable the acquisition of the psychological characteristics possessed by all members of our species, and the age when each appears, are different from those that allow us to distinguish among persons. Most children display their first smile in response to a face, first cry at the sight of a stranger, first imitation, and first full sentence at roughly the same age and in that temporal order. This lawfulness is due, in part, to the maturation of the central nervous system. But the variation from child to child results from a different set of biological processes. The maturing biology that allows all 7-month-old infants to recognize that an approaching adult is unfamiliar, and perhaps to scream, is not the biology that mediates the extraordinary variation in the intensity and frequency of the distress reaction. The differences among young children in the quality and intensity of their emotional reactions and behaviors that are based, in part, on heritable neurobiologies comprise the phenomena we call human temperaments.

Cycles in the Popularity of Temperament

Opinions regarding the relative contributions of biology and experience to behavior have cycled over the years, depending on other ideologies that happen to be ascendant in a given society. Eighteenth-century England, on the cusp of great economic and military power and eager to announce its intellectual separateness from Catholic France and Spain, celebrated the power of the human mind to accomplish whatever it desired without metaphysical restrictions. John Locke satisfied that desire by declaring, without evidence, that each child begins life intellectually innocent, free of both truth and doubt, so that postnatal experiences can create a Chaucer, Shakespeare, or Newton.

A century and a half later, the social price of industrialization was paid in the form of urban ghettos filled with illiterate families that combined high fecundity with equally high mortality. Although some still insisted that Locke was right, Darwin's revolutionary ideas supported those who believed that expensive attempts to alter the status of the poor would not succeed because the wretched children in rags suffering from perpetual colds were inherently different from the rest of society. Hence, it was rational to let them be as they were. Antebellum Americans favoring slavery held a similar view. Thus, social trends led many Western commentators on the human condition to argue that biology won the competition with experience.

History's muse began writing a new chapter when European immigration to America increased during the last decades of the nineteenth and the early decades of the twentieth century. Although

New York ghettos in 1910 resembled the London ghettos of 1850, more egalitarian Americans wanted to believe Locke and the message in George Bernard Shaw's *Pygmalion*. If a poor child were educated properly, he could, like Abraham Lincoln, become president. Although some prominent Americans remained committed to the power of biology, this view could not compete with the ideal of equal dignity for all citizens, and the press began to give headline status to a small number of scientific findings suggesting that experience was sovereign. These reports would have remained unread on dusty library shelves if they had not been useful to the prosecution of an egalitarian ethic.

Thus, for much of the twentieth century, the majority of American social scientists, as well as the general public, were certain that most of the variety in children's moods, talents, and behaviors was the result of social encounters. Biology guaranteed that healthy infants would have a pair of hands, legs, and eyes, but it was largely irrelevant in determining how they were used. Every person given an intact brain and body could actualize her goals if she worked hard and exploited her inherent cleverness; no person's future should be shackled by her biological inheritance.

An important reason why America's first psychologists were not very interested in a child's temperament, even though Jung (1961) and Freud (1950) awarded it influence, was America's celebration of the individual, who, through wit and perseverance, made or invented a reliable product with pragmatic value. Thomas Edison and Henry Ford were prototypical American heroes. In 1836, when the U.S. Congress had to decide what federal office should be built at a central location on the National Mall, they chose the Patent Office. European parliaments would have constructed a cathedral in such a prominent location.

Two of the most celebrated American scientists of this era, the biologist T. H. Morgan and the physicist Robert Millikan, worked alone in laboratories churning out sound facts that promised practical applications. American psychologists interested in how children learned new habits and acquired new skills chose rats, not people, as preferred organisms because these animals were easily bred and had short life spans. The experimental animal laboratory, like a small factory making ball bearings, became the source of reliable knowledge that would improve the community and serve an egalitarian ethos.

Many Europeans at the end of the nineteenth century were loyal to a different ethos. They accepted biological differences among children and celebrated attachment to one's community—village, town, or city—unlike the more geographically mobile Americans, who celebrated the individual. The long-term residents of Paris, London, and Florence felt more pride in their city's history than did those living in New York, Chicago, or Atlanta. Being more concerned with material progress, Americans assumed that if each person did his job well, all would have a happier life, even though their communities might not be as harmonious as they wished.

Thus, the European acceptance of inherent psychological differences, combined with a greater emphasis on the community than the individual, generated an intellectual climate that was friendlier to the concept of temperament. Distaste for biological bases for psychological variation led Americans to reject the idea that some children begin life with a burden they would find difficult to change, while others are born with a gift of cheerful optimism they did not earn.

But history during the second half of the twentieth century began to write a new story with four plot lines. A new immigrant group had arrived from Mexico, Puerto Rico, China, Vietnam, Laos, Ghana, and Nigeria, rather than Germany, Sweden, Ireland, Italy, Poland, France, Latvia, and Russia. These citizens found it more difficult than the Caucasian immigrants of a century earlier to identify with the values of the majority. America needed the labor of the first arrivals; but by 1970, finding a good job in the

American economy and ascending the ladder to middle-class status were more difficult. In addition, the rise of feminism, one of the prizes of the Civil Rights movement, made it easier for young women to pursue any vocation that appealed to them. They did not have to restrict their choice to teacher, secretary, or nurse and could train for medicine, law, business, or one of the sciences if they chose. As a result, some public school systems found it more difficult to recruit women willing to commit their talent and passion to educating young children for all of their working years. The combination of marginalized vocations and a compromised educational experience frustrated youth who wished to better themselves; and movement out of poverty became more difficult. Nonetheless, most Americans remained loyal to the egalitarian promise of benevolent change through better education and liberal government policy.

The distinctive ethical values of the growing proportion of Hispanic, African, and Asian Americans represented a second plot line. Once the moral premises of the Civil Rights movement had become popular, the courts, following the country's changing mood, asserted the equivalent validity of distinct moral positions. Each family's values had inherent dignity. Each individual had a right to believe whatever he wished, as long as it did not interfere with the activities of another. This ideal made it difficult to hold poor families responsible for the academic and vocational failures of their youth. Attempts by government or local institutions to persuade families to encourage academic achievement were viewed by some as an intrusion into a family's privacy. Because it became politically incorrect to blame the victim for his circumstances, the society needed to find something or someone else to blame.

The third plot line supplied the culprit. Quietly, away from the social turmoil, scientists in university and pharmacology laboratories were making extraordinary discoveries involving the influence of genes and biochemistry on brain, mind, and behavior.

Remarkable new scanning machines revealed that the brains of schizophrenic, autistic, dyslexic, and psychopathic patients were different. The media announced these findings to a public eager to explain the social immobility of its recent immigrants in a way that did not require rebuilding neighborhoods and schools and doubling teachers' salaries. The nineteenth-century assumption that many of these children were biologically compromised became attractive again because it eased the guilt generated by the community's failure to improve social and educational conditions. It also shifted some responsibility to drug companies to invent pills that could cure restlessness, retarded language, and impaired attention. The root causes of illiteracy or violence among high school students were attributed to nature's fickleness, over which no one has control.

Changes in the American economy provided the final plot line. Supermarkets replaced corner groceries, twenty-story office buildings replaced small factories, the cost of travel decreased, and Americans who had spent their childhoods in rural Maine, Montana, or Mississippi moved to Los Angeles, Dallas, Chicago, and Seattle. These migrations away from family and friends, together with the new psychology of work, broke whatever communal ties existed before World War I and created a generation of adults trying to gain advantage over colleagues in a ruthlessly competitive society of anonymous neighborhoods and offices.

Empathy toward another person in need is a biological competence that evolution has awarded our species. Hence, humans do not like to be continually wishing misfortune on a peer working in the same office or factory. As a result, Americans were receptive to any scientific evidence that would alleviate doubts about the virtue of their competitive urges and permit them to be self-interested without feeling uneasy. A group of biologists and social scientists answered this need by touting new biological discoveries implying that humans, like other animals, were programmed to be self-interested. Lions kill gazelles in order to survive, and roaming male primates murder young infants in order to impregnate the mothers with their own sperm. There was no need to be ashamed of a desire to put one's own interests above that of friend, colleague, or community because self-interest was nature's intention. And so history returned us to that phase of the cycle in which biology was, once again, regarded as the more important influence on psychological growth.

These events, crowded into the last half century, resurrected the concept of temperament, which had been exiled from American textbooks for almost 50 years. Although we shall see that temperamental biases do contribute to variation in psychological development, this influence is always subject to environmental contingencies. It is rare for two arrows shot with the same force from the same bow to fall at exactly the same place; it is equally rare for two infants with the same temperamental biases, but different experiences, to develop very similar personalities.

The Concept of Temperament

Most scientists define temperament as a biological bias for particular feelings and actions that first appear during infancy or early childhood and are sculpted by environments into a large, but still limited, number of personality traits. The emotional components of a temperament, which many regard as seminal, refer to three qualities: variation in susceptibility to select emotional states, variation in the intensity of those states, and variation in the ability to regulate them. We believe that most temperamental biases are due to heritable variation in neurochemistry or anatomy, although some could be the result of prenatal events that are not strictly genetic. For example, a female embryo developing next to a male will be subject to the masculinizing effect of testosterone from her brother between gestational weeks 8 and 24. As a result, her style of play at 4 years of age will be more similar to that of typical boys than of girls, in being more active and more dominating than the average girl (Collaer and Hines, 1995). Better controlled research with animals reached a similar conclusion.

Because the number of possible heritable brain profiles is much larger than the number of genes, there will necessarily be many temperamental biases, and most remain undiscovered. There are over 290,000 combinations of the 12 known blood types and their variants; hence, the chance of any two people having the same combination is only 3/10,000 (Lewontin, 1995). Because there are many more distinct neurochemical profiles than blood types, the chance of any two people having the same temperamental profile will be far less than 3/10,000. Some temperamental categories, like some blood types, will be very rare, while others will be more common.

Mary Rothbart's bold, synthetic ideas dominate discussions of infant temperaments (Rothbart, 1989; Rothbart et al., 2001, 2003). She posits two primary dimensions on which infants vary—reactivity and self-regulation—and both are controlled by the social environment. The concept of *reactivity* refers to the ease of arousal of motor activity, emotion, and biological systems. *Self-regulation* refers to processes that can either facilitate or inhibit reactivity, especially forms of attention, approach, withdrawal, attack, and restraint, and capacities for self-soothing.

The popular temperamental qualities ascribed to children after the period of infancy resemble some of those posited for infants (Bates, 1989; Buss and Plomin, 1984). *Negative emotionality* refers to the frequent display of distress, fear, and anger. *Difficultness* is a derivative of a category first described by Alexander Thomas and Stella Chess—two New York psychiatrists—that combines irritability, vulnerability to stress, and a demanding posture with parents. *Adaptability to novelty* refers to tendencies to approach unfamiliar events and situations. *Reactivity, activity, attention regulation, sociability,* and *positive reactivity* are five additional temperaments. Although each of these behavioral categories may turn out to be stable and influenced, in part, by biology, the available evidence on human temperaments is most extensive and most persuasive for behaviors that occur in response to unexpected or unfamiliar events.

Reaction to the Unfamiliar

Animals display an extraordinary array of sizes, shapes, colors, coverings, physiologies, diets, and behaviors. But beneath this obvious diversity lies a small number of shared features necessary for coping with each day's challenges. One such property is a readiness to react to an unexpected change in the sensory surround (such as a change in illumination) or to an event that is different from representations created from past experience (a person with an eye patch, for example). Both types of events automatically alert the individual and evoke attempts to relate the unexpected or unfamiliar experience to the present context and recollections of the past. We have been studying the brain states, emotions, and behaviors generated when children encounter the unfamiliar and have come to believe that the profiles we discovered represent two significant temperamental biases.

The brain/mind is always prepared for a particular envelope of probable events for each context in which an animal or human finds itself. The entrance of a parent while a child is playing happily in her bedroom is a highly probable event. But the appearance of a person wearing a clown costume in the same context lies outside the envelope of expected experiences and elicits an automatic stare, immobility, and, in some children, crying—a profile that psychologists call *fear to novelty*. The fact that some children cry intensely and remain immobile for a very long time, while others stare for a few moments and then smile, requires an explanation. We suggest that a temperamental bias for high or low reactivity, detectable during infancy, contributes to this variation. A century earlier this variation was attributed to experience.

As a larger cortical surface evolved in mammals, the range of possible events that could alert the animal expanded. A frog automatically detects a change in the motion of an object and reflexively extends its tongue toward an insect darting in front of its eyes. But the frog is unlikely to do anything if a new frog appears on the other side of the pond. A chimpanzee, on the other hand, immediately notices an unfamiliar animal in its territory and reacts accordingly.

Young children are alerted by unfamiliar people, animals, places, foods, and objects. However, following repeated exposure, and a vocabulary that classifies the novelty, children eventually become accustomed to former discrepancies and cease to look up every time an airplane roars overhead. One frequent class of events, however, continues to provoke an initial vigilant reaction. Children regularly encounter unfamiliar peers and adults whose reactions and evaluations are unpredictable. Parents conceptualize the complementary traits their children display under these circumstances as shyness or sociability, rather than recognizing that their children possess a more general response bias toward all unfamiliar events, not just strangers.

An infant who is easily aroused by unfamiliarity could become one of a large but nonetheless limited number of personality types. The profile that is actualized in adolescence depends on the history of experience. To rephrase Quine, every psychological quality can be likened to a fabric woven from very thin black threads representing biology and very thin white ones representing experience. The cloth appears to be pale gray because the black and white threads are impossible to detect after the weaving is completed.

A temperamental bias can be likened to a genetically programmed brain circuit that permits a large number of motor behaviors. Each culture selects a small number of motor coordinations that are adaptive for its setting. For example, the ability to oppose thumb and index finger allows children in literate cultures to hold a pencil; in nonliterate societies without schools, the same circuit allows individuals to pluck a chicken. Similarly, each temperamental bias is only a potentiality. A child with a temperamental bias for avoiding unfamiliarity who lives in a large city with lots of strangers is likely to become shy. However, in small, isolated villages, where few people are strangers but a variety of animals roam the unpaved trails, this child might become wary upon seeing a three-legged dog or several birds with broken wings.

The reaction to an unfamiliar event depends on how it is perceived (as a threat or harmless), how easily it is assimilated, and whether an appropriate response is available. One-year-olds will reach toward a new toy after playing for a while with a different one because the new object poses no threat, it can be assimilated at once, and a relevant behavior is in the repertoire. However, not all 1-year-olds will reach toward a stranger who has extended her hand because this event is not assimilated quickly and the child is uncertain about the appropriate response. Thus, children, like adults, live in a narrow corridor bordered on one side by the appeal of new experiences and on the other by a fear of novelty that cannot be understood.

Variation in reaction to unfamiliar events is present in mice, rats, wolves, dogs, cows, monkeys, birds, and fish. Pavlov noted, over 75 years ago, that some dogs in his laboratory were unusually tame with humans, while others cowered when an adult made an unexpected movement. Pavlov called the former excitable and the latter inhibited. Observations of five breeds of puppies housed at the secluded Jackson Laboratory in Bar Harbor, Maine, revealed that some breeds were exceptionally timid. A handler would take a puppy from its cage to a common room, place the animal one or two feet away, stand still, and watch. The handler would then slowly turn, walk toward the puppy, squat down, hold out his hand, stroke the puppy, and finally pick the animal up. The puppies who ran to the corner of the room with a high-pitched yelp early in the sequence were classified as timid. Basenjis, terriers, and shelties turned out to be far more timid than beagles and cocker spaniels. But all five breeds were less timid if raised in a home rather than in the laboratory kennels (Scott and Fuller, 1965).

Selective breeding of a strain of fearful or fearless animal off-spring requires a relatively small number of generations. For example, some quail chicks become immobile when placed on their back and restricted by a human hand; the immobility is regarded by behavioral biologists as an index of fear. Fewer than 10 generations of selective breeding were needed to produce a relatively homogeneous line of chicks that became immobile when placed on their backs and restrained (Williamson et al., 2003). It is tempting to assume that the biology which mediates the avoidance of unfamiliarity is essentially the same across species. But it is possible that the bases for intraspecific variation in dogs, cats, monkeys, or birds are unique to that species. That is, the biology responsible for a timid cat may not be the biology responsible for a timid monkey or child.

Preservation of Behavior in Response to the Unfamiliar

When teachers in kindergarten and, later, in sixth grade rated over 1,800 Canadian children for fearfulness, most of the sixth-graders received a rating that resembled the one they had received in kindergarten (Cote et al., 2002). Shy, over-controlled Icelandic children were more likely than their classmates to remain shy as adolescents (Hart et al., 1997). Two-year-old children who were described by their adoptive Dutch mothers as shy and dysphoric possessed similar traits when they were 7 years old. Neither the family's social class nor the mother's sensitivity with her infant predicted these psychological properties at 7 years (Stams, Juffer, and van IJzendoorn, 2002; Rimm-Kaufman et al., 2002).

Behavioral observations of a large sample of identical and fraternal twins observed at 14, 20, 24, and 36 months revealed that inhibited or uninhibited behavior in response to unfamiliar events had a heritable component (Robinson et al., 1992; Kagan and Saudino, 2001). The inhibited children remained close to their mother and avoided playing with unfamiliar toys and peers; the uninhibited children showed the opposite behaviors. Heritability estimates were higher when the sample was restricted to children who were extremely inhibited or uninhibited in a play session with two other unfamiliar children (DiLalla, Kagan, and Reznick, 1994). (For a critique of heritability estimates, see Chapter 6, below.) But parental behaviors affected the degree of preservation of these traits. Inhibited 2-year-olds were most likely to preserve their style if they had intrusive, hypercritical mothers and were less likely to remain reticent with peers if their mothers discouraged a shy posture (Rubin, Burgess, and Hastings, 2002).

A temperamental bias renders some children vulnerable to a serious bout of anxiety following a traumatic experience. For example, 10 of 40 California school children who were kidnapped and terrorized for two days developed post-traumatic stress disorder (Terr, 1979). During the winter of 1984, a sniper fired at a group of children on the playground of a Los Angeles elementary school, killing one child and injuring 13. One month later, when clinicians interviewed the children to determine who was experiencing extreme levels of anxiety, 38 percent were judged to be anxious, while 39 percent seemed free of unusual levels of tension or fear. The children who were judged to be anxious had shown an inhibited style prior to the school violence (Pynoos et al., 1987).

A group of fourth- and fifth-grade children living in south Florida had been assessed for the presence of anxiety 15 months before Hurricane Andrew struck the area. The 11 percent of the children who remained distressed 7 months after the fierce storm had been categorized as anxious prior to the hurricane (La Greca et al., 1998). Similarly, during the bombing of London in World War II, British children under 5 years of age who became fearful after being taken from their homes for their own safety had shown signs of being highly fearful before the bombing raids began (John, 1941).

Freud's reluctance to acknowledge that some children had a temperamental vulnerability to be fearful of unfamiliar events is puzzling because this idea was prevalent in the professional literature with which he was familiar. Surely Freud saw some children clutch their mother's hand when a stranger approached. Freud confronted this issue in *Inhibitions, Symptoms and Anxiety* (Freud, 1926). After noting that young children are afraid when alone, in the dark, or with someone they do not know, he anticipated John Bowlby by arguing that "these three instances can be reduced to a single condition; namely, that of missing someone who is loved and longed for." Freud then exploited a semantic trick by suggesting that fear of losing a loved one resembled castration anxiety because both states involved the loss of something valuable. Freud hoped that his readers would regard the two states as identical simply because the notion of "losing" a valued resource was present in both statements.

This semantic permissiveness is common. The number of distinct verb forms is much smaller than the number of distinct nouns that a predicate serves, especially in English. One unfortunate consequence is that readers are seduced into believing that a predicate which refers to a behavior under one set of circumstances retains the same meaning when it is linked with very different agents. The meanings of the word *bit* in "The rat bit an intruder" and "The boy bit his sandwich" are dissimilar, as are the meanings of *lost* in "The child lost his confidence" and "The child lost his pencil."

Freud wanted to dismiss unfamiliarity as the source of a temporary state of uncertainty because he was concerned with chronic anxiety rather than the acute state evoked by an unfamiliar person or place which vanishes after several minutes. Most adults are afraid of neither strangers nor new places; hence, it was easy to conclude that a fear reaction to unfamiliarity was a transient property of childhood. Finally, Freud's desire to be loyal to parsimony required him to posit only one form of anxiety. The aesthetic appeal of his ideas would have been diluted if he had nominated several different types of anxiety provoked by different origins. A similar aesthetic standard motivated the Greeks to assume only four temperamental types to match the number of combinations of the qualities warm-cold and dry-moist. Freud failed to heed Francis Bacon's warning, "Let every student of nature take this as his rule, that whatever the mind seizes upon with particular satisfaction is to be held in suspicion."

We believe that the temperamental biases we call high- and low-reactive in infancy and inhibited and uninhibited in children are due, in part, to distinct neurochemical profiles in the amygdala and its projections to cortex, basal ganglia, and autonomic and endocrine systems following encounter with unfamiliar events (Knight, 1996; Rolls et al., 2003). Even though each temperament emerges from the combination of an initial biological bias and later experience, the exact nature of that relation remains obscure.

Impediments to Understanding

The first impediment to a deeper understanding of the relation between biological and psychological processes is semantic, for the descriptions of biological and psychological phenomena use different vocabularies. The language of neuroscience contains references to neurons, receptors, circuits, neurotransmitters, and their functions (for example, excite, inhibit, bind, secrete). The language of psychology refers to schemata, semantic networks, feelings, emotions, motor responses, and their functions (such as acquire, retrieve, recognize, display, experience). The psychological statement, "The 1-year-old infant stopped crying to the approach of an unfamiliar man with a beard when she saw her mother smile," cannot, at the moment, be translated into sentences that contain only the language of neurobiology.

Even if scientists could measure the brain states occurring during the 10-second event noted above, the biological evidence would not allow them to know how the infant behaved without knowledge of the context and the infant's history. This observation is neither a defense of mind/brain duality nor a critique of biological reduction. It simply asserts that each perception, inference, emotion, and behavior emerges from, and therefore is more than, the brain circuits necessary for its actualization—a position Roger Sperry promoted during the final years of his productive career—because most behaviors can originate in different brain states. Just as a wave crashing on a beach is not equivalent simply to its billions of water molecules, a temperament is not equivalent to the biological phenomena that are its foundation.

A second impediment to appreciating the relation between biological events and the feelings and responses that define a temperamental bias is methodological. The tools available to neuroscientists permit only a partial evaluation of the rapidly changing, complex patterns of brain activity that are the bases for psychological phenomena. Consider an example. The sight of a dangerous snake 10 yards away activates sensory, limbic, frontal, and motor cortical sites that lead an individual to flee in haste and, later, to report feeling fearful. This psychological event, from the first sight of the snake to the first movement of the legs in flight, takes less than a second. Since the brain's reaction to a snake on a trail cannot be assessed with an fMRI scanner, scientists simulate this reaction by putting people in a scanner and showing them pictures of snakes. But photos of snakes viewed in a laboratory and real snakes encountered in the forest are very different events. It is unlikely that the pattern of brain activity recorded in a scanner would resembles the pattern recorded on the forest trail, if such measurements were possible.

Similarly, the pattern of brain activity in a person watching an erotic movie while lying in a PET or fMRI scanner would be different from the pattern evoked when the same person is watching the same film alongside a lover in a hotel room. Neuroscientists may use the term *sexual arousal* to describe the person's state in both situations, but the brain patterns are likely to be distinct. As Niels Bohr noted almost a century ago, scientists can never know a phenomenon as it exists in nature; all they can ever know is what they can measure (Bohr, 1933). A drop of black ink disappears if it is stirred in a cylinder of glycerin. Although it is factually correct to state that the cylinder contains black ink, the latter is invisible. So, too, for human temperaments. A young infant's temperamental bias for vigorous activity and distress in response to unfamiliar events is embedded in a family context that, over time, creates a psychological profile. The temperamental contribution is an intimate part of that profile, but it cannot, at the present time, be detected as a separate component, even though scientists write about it that way.

What Can We Measure?

Most important advances in biology have been the result of scientists first discovering and then probing a surprising new phenomenon, rather than beginning with an a priori notion of what they believe to be true and searching for its instantiation in observations. Both Darwin and Wallace had to travel to places very different from their native England to see for themselves the phenotypic variety in animals believed to belong to the same species. Both men could have imagined this variety while sitting in their studies at home, but they failed to do so. Darwin's insights required his firsthand observation of the variation among finches and tortoises; Cajal's generative ideas came from microscopic study of brain tissue; Watson and Crick had to see Rosalind Franklin's X-ray photo of DNA in order to infer the correct structure of the molecule. Had Pavlov first decided to study the nature of human learning, he probably would not have conditioned salivation in dogs.

We faced a similar set of options during our initial inquiry into temperamental biases. We could have been bold and invented an a priori set of temperamental properties and their accompanying behavioral and biological features. But, being suspicious of heady theory in this immature domain of inquiry, we chose the more conservative strategy of first selecting behavioral profiles we had learned were preserved over time, and then studying their development and biological features more carefully. In 1980 and today, the evidence seemed to us too weak to support a declaration, based on one or more a priori principles, that certain qualities were temperamental.

Some readers may wonder why we were cautious. Thirty years earlier, Thomas and Chess (1963) had interviewed parents and proposed that some infants were slow to warm up to strangers while others were sociable. These categories resemble our distinction between inhibited and uninhibited children. But the parents Thomas and Chess interviewed did not describe the properties of high- and low-reactive infants. Scientists who used parental questionnaires chose ease of arousal and regulation, treated as continuous traits, as primary concepts (Rothbart, 1989) and were indifferent to the form of the arousal—crying, smiling, thrashing limbs, or babbling. Had we accepted these popular ideas, we would have missed the critical fact that, in response to unfamiliar stimuli, high-reactive infants combine vigorous motor activity with frequent crying, rather than with babbling or smiling. Low-reactive infants often babble and smile, but they rarely cry or thrash their limbs.

We had to see the variation in the combination of motor activity and crying when children were presented with colorful mobiles, voices, and smells in order to infer, a posteriori, the concepts of high and low reactivity. We suspect that we would not have generated these constructs had we sat at a desk thinking about the theoretically most fruitful ways to classify infant temperaments, or had merely asked parents to describe their infants.

Continua or Categories

We also had to decide whether the behaviors defining a temperament should be considered as continuous dimensions or whether a certain range of values represented a discrete temperamental category. This issue remains a source of healthy debate. The unaided perception of nature reveals many clearly bounded objects and events—a tree, cow, and snow storm. However, the variation within the categories trees, cows, and snow storms motivates the mind to invent concepts for the events that share essential features.

Even though scientists recognize that the category names are arbitrary, they would like to know whether all the objects within one category should be viewed as forming a continuum or whether some have features sufficiently distinct to set them apart from other members. Should a flowering dogwood, for example, be regarded as on a continuum with a 100-foot spruce or as a qualitatively different kind of tree? Is a 10-minute snow shower qualitatively different from, or on a continuum with, a 12-hour blizzard?

Most biologists favor categories over continua because life forms are considerably more varied than sand, stones, and stars. All photons are identical; all bacteria are not. Assume that two fraternal twin sisters differ in only 10 alleles that influence eye color, and that one girl has brown eyes and her sister blue eyes. The investigator must decide whether to emphasize the fact that the sisters share 0.9999 percent of their genes and place them on a continuum, or focus on their different eye colors and classify them as belonging to qualitatively different groups. Either position is defensible; the one chosen depends on the theoretical purposes of the investigator.

If blue- and brown-eyed children have different access to resources, or are differentially prone to certain illnesses (it turns out that blue-eyed individuals are more prone to melanoma), an investigator might be tempted to argue that the girls belong to distinct categories. If, on the other hand, their distinct features have no known implications, a scientist might place the girls on a series of continua. This issue divided Karl Pearson and George Yule—two important founders of modern statistical methods. The former saw continuous distributions, while the latter saw the utility of categories.

The current preference among psychologists for continua over categories also derives from the contagion of ideas across disciplines. Before the introduction of relativity theory about a century ago, physicists assumed that mass and energy were qualitatively different. A burning log was distinct from the energy contained in the heat and light it emitted. But if mass and energy are interchangeable, psychologists could defend the notion that no person was qualitatively different from another on any psychological property; all could be located on a set of continua.

A majority of psychologists prefer to treat temperaments as continuous traits because the most popular statistical techniques require this approach. The use of inferential statistics became the mark of the well-trained social scientist after World War II, and psychologists realized that their technical papers were more likely to be accepted if they used correlation coefficients, t-tests, and analysis of variance and covariance to analyze their data. Faculty members trained their graduate students to treat people as substantially similar in their sensations, perceptions, memories, and emotions so that statistical analyses could be performed on continuous scores produced by different experimental conditions rather than by different kinds of people.

The pressure to restrict statistical manipulations to analyses of variance and regression has frustrated those who intuit that some subjects are qualitatively different from the majority in their sample. But, unfortunately, there is no consensus regarding the algorithms that permit an investigator to conclude that some children belong to a distinct group. Consider an investigator, unfamiliar with Down syndrome, studying the relation of maternal age to children's intelligence in a sample of 600 families. The correlation between maternal age and IQ would reveal no statistically significant relation. However, an examination of a scatterplot would reveal that two children with very low IQ scores had the two oldest mothers in the sample. Reflection on that fact might tempt the investigator to conclude that these two children are qualitatively different from the other 598 and that these two families provide a clue to a relation between the age of the mother and the child's intelligence for a very small proportion of the population.

One child from the sample of over 500 that we studied showed a unique developmental profile. When he was examined at 8 weeks

of age, this boy frowned spontaneously 6 times either during or between a variety of stimulus presentations. Spontaneous frowns are
rare among infants and especially rare during the quiet interval between stimuli. This boy continued to frown frequently in response
to a similar battery when he was 4 months old and retained a very
sad facial expression for periods as long as 30 seconds. This combination is rare. The boy emitted a sharp scream when presented
with a moving mobile, and when he was observed at 9 months he
cried at the presentation of every stimulus.

This child continued to be idiosyncratic when he was observed
at 14 months. His face showed sadness throughout the 90-minute
battery, he displayed several uncontrolled tantrums, refused most
episodes by screaming, and often displayed a pained expression of
angst, without an accompanying cry. His mother acknowledged to
the examiner that he had become aggressive lately and noted, "He
walks up and bites you."

When he was 3.5 years old he was calmer and showed no unusual behavior during the 90-minute session. However, several
weeks later when he was being observed at play with an unfamiliar
boy of the same age, the boy left his mother after 5 minutes of
play, went to the center of the playroom, and began to punch a
large inflated toy. Two minutes later, he seized the toy that the
other boy was holding and retreated to his mother. Five minutes
later, when the other boy was inside a plastic tunnel, he picked up
a wooden pole and began to strike the tunnel in the place where
the other boy was sitting. The force of the blow made the other
boy cry. This unprovoked act of aggression toward an unfamiliar
child is an extremely rare event in a laboratory context with both
mothers present in the room.

This boy showed unique reactions on almost every assessment.
Most psychologists observing this child for the first time at 3.5
years of age would probably dismiss the single act of aggression as
reflecting greater than normal frustration on that particular day.
However, in light of his history, we suspect that the impulsive act

reflects a deep psychological quality, and this boy is a member of a rare temperamental category.

Scientists who study temperaments have to choose between continua and categories when they interpret their evidence. Imagine that 10 genes determine the density of receptors for the neurotransmitter GABA in the basolateral nucleus of the amygdala (about which more will be said in Chapter 3), and only the children who inherit all 10 variations of these alleles have a very low density of GABA receptors in this site. These children should have a low threshold for distress in response to novelty and show extreme responses to unfamiliar events. Although some investigators might argue that distress and fearfulness are continuous variables and therefore deny categorical status to these children, it is equally reasonable to argue that the children with all 10 alleles represent a qualitatively distinct group.

The heights of 10,000 children form a continuous, almost perfectly bell-shaped distribution. Hence, it is tempting to assume that the root causes of variation in height differ only quantitatively. We know this assumption is false. Some individuals inherit a particular genome that makes it likely they will be very short in most, but not all, environments, even though seriously malnourished children will be equally short. Thus, a small proportion of very short children have a unique biology.

Many functional relations in biology and psychology are nonlinear. When a process attains a certain magnitude, novel mechanisms emerge that change the form of the function. For example, the magnitude of startle in a rat is small if the intensity of electric shock used during conditioning is very weak or very intense. Startle is maximal if the shock is of moderate intensity. We often found no significant differences between two temperamental groups when we compared mean values, but we occasionally discovered significant and theoretically meaningful differences between high- and low-reactive children when we compared scores in the top and bottom quartiles. If temperaments are categories, comparisons of children with extreme scores on a measure are appropriate. A biologist phrased the case for qualitative categories: "The study of biological form begins to take us in the direction of the science of qualities that is not an alternative to, but complements and extends, the science of quantities" (Goodwin, 1994, p. 198).

However, conceptualization of psychological or biological variables as continua or as categories depends, in the end, on the theoretical interests of the investigator and the methods chosen to serve those intentions. If the primary concern is with predicting academic performance, IQ can be treated as a continuous distribution. Bur if the wish is to understand the contribution of genetic factors to very low IQ scores, it is useful to treat certain IQ ranges as categories because the genetics of Down syndrome differ from the genetics of phenylketonuria (PKU). Thus, nature permits scientists to favor both strategies.

We view temperaments as categories because we believe that the origin of each is a distinct genomic profile. Therefore, children who are extremely inhibited or uninhibited in response to unfamiliarity are qualitatively different from those with intermediate levels of shy, avoidant behavior (Woodward et al., 2000). This view is supported by both theory and evidence, as later chapters will show. Consider one source of support for this position. Rubin and colleagues found that only 10 percent of a sample of 2-year-olds were consistently inhibited across varied unfamiliar events, and only the consistently inhibited children had low vagal tone (Rubin et al., 1997). These investigators also distinguish between autonomous children who play alone but are not excessively wary and those who play alone because of uncertainty. Only the latter group showed biological signs of arousal (Henderson et al., 2002; see also Reznick et al., 1989).

Self-Reports

One reason many psychologists regard temperaments as continua is the fact that, at present, questionnaires are the primary source of evidence. Although questionnaires asking parents to describe their children are the current method of choice, this strategy has flaws. Investigators who rely only on questionnaires assume that a person's semantic judgments about their behavior, beliefs, and moods, or those of their children, are valid bases for inferring fundamental psychological categories. This claim is surprising. No immunologist would use reports of informants as the basis for determining distinct categories of diseases; no economist would rely on interviews with consumers to formulate the fundamental concepts of economics; no cognitive psychologist would restrict his inquiry to adult descriptions of thoughts and perceptions in order to infer basic cognitive abilities.

Many problems trail the use of questionnaires when they are used as the only basis for inferring temperamental, or personality, categories. Two critical functions of mind mediated by distinct brain circuits are recognition of the familiar features of an event and retrieval of representations of the past. These functions involve the psychological structures we call schemata and semantic forms. The former represent perceptual events; the latter linguistic ones. Although these two structures are usually intertwined in recognition and retrieval, each preserves its distinctive properties, as do sodium and chlorine when they combine to make salt. Each semantic network varies with respect to its connections to schemata. The semantic networks for *girl, apple,* and *table* are more richly plaited with schemata than the networks for *surtax, truth,* and *knowledge.* The meaning of a word or a sentence is contained in the pattern of relations among the combined semantic and schematic forms.

Because questionnaires rely on semantic networks, respondents vary in the particular schema that are linked to the words in the questionnaire item; hence individuals do not always extract the same meaning from a question. A person who activated schemata representing loss of behavioral control and display of an aggressive act as he read "Are you easily angered by frustrations?" might answer "No," while one who activated schemata for a private feeling of tension might answer "Yes" to the same question.

Second, most people try to give semantically consistent answers, and this motive leads to distortion. A woman who affirms on one question that she likes meeting new people will be biased to respond affirmatively to all related questions in order to maintain semantic consistency. Terms like sociable and shy are antonyms; therefore, the features linked to each of these words are inversely correlated in the semantic networks of respondents, even though many adults who like meeting new people take equivalent pleasure from periods of reading or hiking alone. But the semantic opposition of outgoing and solitary biases the respondent to minimize one or the other of those two dispositions. Most parents treat the concepts happy and sad as antonyms. Therefore, mothers who believe their infants laugh frequently will resist describing them as irritable, even though films of infants reveal that a large group of infants both laugh and cry frequently.

Third, if a human trait does not have a popular name and therefore is not part of a semantic network, a questionnaire often omits relevant items. Some men are loyal to their wives and affectionate with their children but disloyal and hostile in their relationships with colleagues. There is no current semantic concept that reflects that constellation of traits. In addition, sentences on questionnaires usually refer to categories of behavior or mood rather than blends. Languages are not rich enough to describe all the schematic experiences that are possible, and those who design questionnaires must choose one query from a much longer list.

Verbal descriptions are too coarse a measure of emotions and moods because the vocabulary available to most informants is too limited to capture subtle differences in feeling states that could have theoretical significance (Schnitzler et al., 1999). Shy adolescents vary in the quality of their conscious feelings when they interact with strangers, but the words available to them are limited to terms like *tense, nervous, worried,* or *anxious.*

Human languages evolved primarily to serve thought and communication. Language permits one person to tell another about the location of food or predators and to instruct or remind people of their social responsibilities. Humans usually try to disguise their emotions, especially envy, anger, shame, guilt, sexual arousal, and conflicted intentions to deceive. More important, most people, in describing their emotional state to another, often include a reference to the setting that provoked the emotion. Because the members of a language community understand that the quality of a person's feeling depends on the specific context, there is no need to invent a new emotional term for each class of context.

For example, the quality of an adolescent's feeling following an ethical lapse varies with the exact nature of the lapse and the social context in which it occurred. An adolescent informs listeners of these facts when he says to a friend: "I felt guilty when I failed my teammates by playing poorly in the soccer game." This specification of the act and its context allows listeners to know that the quality of the adolescent's emotion in this situation probably differs from what it would be if he had said that he felt guilty for not giving his younger brother an extra piece of cake. No questionnaire designed to measure variation in emotional experience describes in sufficient detail the combination of action and context. Further, questionnaires cannot ask about all the situations that might be incentives for emotional reactions in particular persons. For example, some adolescents worry a great deal over the health of a close relative; others worry most over their personal appearance; and still others about their grades in school. This specific knowledge might illuminate the psychology of these three types of adolescents.

Fourth, variation in biological activity, which influences the quality and intensity of moods and feelings, lies beyond the reach of a questionnaire because individuals do not have conscious access to this activity. Two people could report equivalent anxiety over their health but differ in degree of activity in the HPA axis and therefore in the quality and intensity of their anxiety as perceived by an intimate friend. Caucasian adolescents who had very high scores on a questionnaire index of anxiety sensitivity were at high risk for a panic attack, but Asian and Hispanic adolescents with equally high anxiety sensitivity scores were not at special risk (Weems et al., 2002). Adults with social phobia and those with facial disfigurement report the same level of anxiety over interacting with others, but only the former are reluctant to leave their homes (Newell and Marks, 2000). Patients with spinal cord injuries, who experience less sensory feedback from their body and therefore less intense change in somatic sensations than others, report verbally the same level of psychological arousal to unpleasant pictures as do normals (Cobos et al., 2002).

Although the brain reactions of men and women following intravenous administration of procaine—an anesthetic that affects limbic sites—were different, the two sexes reported very similar descriptions of their feelings (Adinoff et al., 2003). The fact that adults with relatively high blood pressure and heart rate report more frequent daily stressors does not necessarily mean that they actually encounter more stress, for their heightened sympathetic reactivity may make them more aware of their unpleasant feelings (Carels, Blumenthal, and Sherwood, 2000). This is one reason why questionnaire reports of depressed feelings by college students should not be equated with a professional diagnosis of depression; only psychiatrically depressed patients show biological features.

Fifth, reports of the frequency and intensity of past emotions are always influenced by a comparison of the present with the past. However, a verbal report of how one felt at parties attended several years earlier is suspect because of the poor accuracy with which people recall past feelings and actions. One investigator interviewed 67 adults who had been interviewed 30 years earlier when they were in high school. The adults' answers to the same questions posed 30 years earlier bore little relation to what they had said as teenagers.

These five problems are exacerbated when parents are asked to describe their children. Mothers who never attended college typically describe their infants as more difficult to care for than do college-educated parents. Mothers experiencing intense stress have a lower tolerance for frustration and tend to exaggerate their infants' irritability. As a result, the agreement between parents and observers describing the same infants is usually low (Kochanska et al., 1998). For example, parents and observers were consistent in their independent weekly evaluations of an infant's mood from 4 to 6 months of age. But the parents and observers wrote very different descriptions. The authors concluded, "The most important implication of our findings is a cautionary message. Mothers are a poor source of information about their infant's behavioral style" (Seifer, Sameroff, and Krafchuk, 1994).

One well-designed study evaluated the relation between parents' ratings of an inhibited profile in their preschool-aged children and independent behavioral observations of those same children in a laboratory. The mothers' descriptions of their children's shyness around strangers agreed with observers only for children who were extremely shy or extremely sociable in the laboratory setting. The maternal descriptions bore little relation to behavior in the laboratory for over 80 percent of the sample; the mothers of many children who were quiet and subdued with a stranger in the laboratory described them as very sociable (Bishop, Spence, and McDonald, 2003).

A person's answer to a question as simple as "Are you happy?" can be influenced by the form of the question as well as the content of prior questions. When preadolescents were asked to name the events they feared most, spiders topped the list. But when they were given a long list of potentially fearful events, spiders were checked far less often than events like "being unable to breathe" and "burglars." Thus, the way self-reports are gathered affects the scientist's conclusions (Muris et al., 2000). On some occasions, questionnaire evidence leads to conclusions that violate both biology and common sense. One team of investigators interviewed 794 pairs of adult female twins about their physical health and emotional states. The replies to the questions posed by a stranger revealed the surprising fact that "self-esteem" was as heritable as physical health (Kendler, Myers, and Neale, 2000). Had the evidence consisted of a physical examination with laboratory tests of blood and urine together with direct observations of behavior in varied settings, the results would have been very different because the meaning of self-esteem and physical health would have changed dramatically. One psychologist summarized the problem this way: "Retrospective behavior reports are highly fallible and strongly affected by the research instrument used" (Schwarz, 1999, p. 100).

Psychologists who view questionnaires as sensitive measurement devices fail to recognize that many respondents reading an item think about the content of that item for the first time and often arrive at an answer they had not conceptualized prior to the assessment. For all of these reasons, we suggest that conclusions about human temperaments based only on questionnaires or interviews resemble Ptolemy's conclusions about the cosmos based on staring at the sky without a telescope.

The seminal point is that semantic and schematic representations of one's experiences have different structures and award salience to different features. Consider a 2-year-old who usually resists his mother's requests, often with a smile, until the parent threatens a mild punishment, after which the child conforms. These sequences, which typically last several minutes and have occurred many hundreds of times, are represented as perceptual structures in the parent's mind. The mother's reply to a request to rate her child's obedience on a 5-point scale depends on which events she retrieves. Thus, a parent could give different answers on different occasions to the same question. This outcome is especially likely if the parent had not conceptualized the child's typical behavior with the semantic terms *obedient* or *disobedient*. Each parent is forced to condense into a single semantic judgment many hundreds of interactions with her child over the past months or years. This is a burden that language does not bear well.

We are not the first to criticize the validity of parents' descriptions of their children. Over 65 years ago a team of psychologists noted the poor relation between what actually happened during an infant's first year and the mother's recall of those early events when their children were 21 months old (Pyles, Stolz, and MacFarlane, 1935). It is even possible that the words parents use to describe their children, compared with the categories scientists use when they code a frame-by-frame analysis of behavior, may be incommensurable because many scientific terms do not exist in the parents' vocabulary. If asking parents about their children were an accurate source of information, the field of personality development would be one of psychology's most advanced domains. Wise people have been observing children and constructing theories of development for a very long time. The immaturity of our understanding implies that verbal statements describing a child's emotions and behaviors have some, but limited, validity.

History continually casts aside once popular methods in all the sciences. Archeologists now use carbon dating to establish the age of a fossil; evolutionary biologists use blood proteins to assign an animal to a species. The progress from Mendel to the cloning of a sheep rests on the invention of new methods. The recognition that sole reliance on parents' verbal reports does not provide a sufficiently sensitive index of a child's temperamental profile should be regarded as scientific progress. We do not suggest that questionnaires have no value—we used parental reports in our assessment—only that this form of evidence should be supplemented with direct behavioral observations and, when possible, with biological evidence. Our reasons for including biological variables in our assessment are defended in the next chapter.

3

BIOLOGICAL RESPONSES
TO UNFAMILIARITY

We suspect that most behaviors that are derived from temperamental biases originate in heritable neurochemical profiles, an idea anticipated over 75 years ago (Rich, 1928; McDougall, 1929). The mating behaviors of prairie and montane voles—two closely related strains that resemble mice—provide an unusually convincing illustration of the cascade from genes to brain chemistry to behavior. Prairie voles form stable pair-bonds following an initial mating, while montane voles do not. Although both strains secrete vasopressin and oxytocin and both have receptors for these molecules, they vary in the promoter region of the genes that influence distribution of the receptors for these molecules. Prairie voles have a denser set of receptors in limbic sites believed to mediate states of pleasure (Insel, Wang, and Ferris, 1994).

The human brain contains over 150 different molecules that, along with their receptors, influence the excitability of neurons and circuits. The number of possible combinations of these molecule-receptor patterns exceeds many billion; therefore the number of brain profiles that can influence behavior exceeds by a substantial amount the number of classes of behavior. Some molecules are excitatory; some are inhibitory; and the density of receptors for a molecule can be independent of the concentrations of that molecule in each site. These molecule-receptor combinations can affect the excitability of a neuronal ensemble in four ways, and the balance among them contributes to the brain state at any particular moment.

- The ensemble can secrete a greater amount of a particular molecule.
- The ensemble can have more receptors for a molecule.
- Another ensemble that projects to the one of interest can secrete more of a particular molecule.
- An ensemble that modulates the ensemble of interest can be inhibited by a third ensemble.

The molecules that, at present, appear to have the greatest significance for the temperaments include serotonin, GABA, glutamate, dopamine, norepinephrine, opioids, corticotropin-releasing hormone, oxytocin, vasopressin, acetylcholine, prolactin, and the sex hormones.

Illustrative Neurochemical Profiles

The neurotransmitter serotonin is among the best studied of the molecules affecting temperament. Individuals with low levels of brain serotonin are more susceptible to bouts of anger, depression, or fear. Drugs that slow down the re-uptake of serotonin in the synapse, leaving this molecule available for a longer period of time, help a large proportion of depressed and anxious patients. Prozac and its many relatives (the SSRIs) belong to this category.

GABA, like serotonin, inhibits neuronal excitement, while glutamate is excitatory. Adult female bulls born to aggressive or nonaggressive breeds differ in the balance between GABA and glutamate in the pons, cerebellum, and ventral tegmentum, with the aggressive bulls having lower GABA and higher glutamate levels (Munoz-Blanco and Castillo, 1987). The therapeutic effects of the drugs called benzodiazepenes (Valium and its relatives) are due in part to their influence on GABA receptors. A molecule called gastrin-releasing peptide acts on receptors located on interneurons in the lateral nucleus of the amygdala, causing them to release GABA which, in turn, inhibits neural activity. It is relevant that mice lacking the gene for this receptor fail to release GABA within the amygdala and, as a result, preserve traces of associations between a conditioned stimulus and electric shock for a longer period of time (Shumyatsky et al., 2002).

The speculation that human infants born with a compromise in GABA or serotonin function should have difficulty modulating extreme states of distress has some empirical support. Unusually irritable 2-year-old Caucasian children, compared with relaxed toddlers, inherit a shorter allele in the promoter region for the serotonin transporter gene (Auerbach et al., 1999). Japanese infants are less irritable than Caucasian infants, and Japanese adults are more likely than Europeans to have the longer version of the allele (Kumakiri et al., 1999). In addition, adults with the shorter version of the allele display greater activity in the amygdala to fear-inducing events than adults with the longer form (Hariri et al., 2002). However, the picture is not as consistent as scientists would wish, for very shy Israeli children inherit the longer form of the allele for the serotonin transporter gene (Arbelle et al., 2003).

The neurotransmitter dopamine and its many receptors in varied sites affect an animal's tendency to approach the unfamiliar. Most animals initially avoid a novel place, food, or object. However, the restraint wanes if no aversive experience follows, and the animal eventually approaches the unfamiliar. The moment a rat places his forepaws into a novel environment, there is an immediate release of dopamine in the nucleus accumbens that lasts for about 8 seconds (Rebec et al., 1997a,b; Bevins, 2001; Bevins et al., 2002; Schultz, 2002). Rats explored, over a series of days, a distinctive compartment that contained a different unfamiliar object each day (for example, a tennis ball, block, or piece of styrofoam). When the animals subsequently had to choose between entering the compartment that had contained the novel objects or entering a distinctly different place, they chose the former. Because animals given a drug that blocks one of the dopamine receptors failed to choose the compartment with the novel objects, it is reasonable to conclude that dopamine plays a role in the brain state created by unfamiliar events (Bevins et al., 2002).

Dopamine-producing neurons in the ventral tegmentum show an immediate increase in activity when an animal encounters a desired event (often food) it does not expect. A rat trained to receive a certain amount of food in a particular place eats more than usual if it finds more food than it expected (Roitman et al., 2001) or if the time of day that the laboratory lights are turned off is altered and the laboratory becomes dark earlier in the 24-hour cycle. However, this reaction habituates quickly once the food reward has lost its novelty, and dopamine release fails to occur if the reward is expected (Schultz, 2002). It is likely that variation in dopamine levels, or in any of the dopamine receptors, has similar consequences in humans.

Variation in norepinephrine and its varied receptors also affects the response to novelty and, in addition, alertness, the ability to sustain attention despite distraction, and thresholds for detecting subtle changes in sensory signals. It is easy to invent an argument that relates this variation to temperamental qualities (Cecchi et al., 2002). The amygdala, which is modulated by norepinephrine, sends projections to the nucleus accumbens. Hence, greater norepinephrine release in the amygdala should be accompanied by a more excitable nucleus accumbens. Rats from the Wistar strain who usually explore unfamiliar areas have greater norepinephrine activity in the nucleus accumbens than animals who are less exploratory (Roozendaal and Cools, 1994).

Opioids modulate level of excitation in the brain and autonomic nervous system, and the brain sites that mediate pain and subjective feeling tone contain receptors for opioids. The medulla is an important way station for downward projections from the sympathetic nervous system as well as ascending projections from the viscera and muscles to the amygdala. Possession of a high density of opioid receptors, or a high concentration of opioids in the medulla, would mute impulses arriving at medullary sites. Hence, children with greater opioid activity in one or more areas of the medulla might experience more frequent moments of serenity, while those with less opioid activity would be vulnerable to more frequent bouts of tension (Miyawaki, Goodchild, and Pilowsky, 2002; Wang and Wessendorf, 2002).

Not all of the variation in opioid activity is genetic; some originates in prenatal events. For example, a female mouse embryo can lie between two males, next to a male, or surrounded by two females in the uterine horns. The female embryos in the first two positions, who will be affected by the surge in testosterone secreted by the nearby male, develop a greater density of opioid receptors in the midbrain and, after birth, have a higher threshold of tolerance for levels of intense heat (Morley-Fletcher et al., 2003).

Corticotropin-releasing hormone (CRH), produced primarily in the paraventricular nucleus of the hypothalamus, is another molecule with diverse excitatory influences. CRH influences many systems, but especially the hypothalamic-pituitary-adrenal axis (HPA). One product of activity in this axis is secretion of the hormone cortisol by the adrenal cortex. Capuchin monkeys with high cortisol levels are more avoidant than animals of the same species with lower levels (Byrne and Suomi, 2002), and the gregarious bonnet macaque has lower concentrations of CRH than the highly volatile, less gregarious pigtail macaque (Rosenblum et al., 2002). Further, rats who fail to explore an unfamiliar area show a greater increase in corticosterone than more exploratory animals (Cavigelli and McClintock, 2003). First-grade children who were classified as fearful showed their highest cortisol levels in the evening, while fearless, more sociable children displayed higher cortisol levels in the morning (Bruce, Davis, and Gunnar, 2002). There is even evidence for a relation between an allele at a CRH-linked locus and behavioral inhibition in children whose parent, or parents, had panic disorder (Smoller et al., 2003).

However, cortisol secretion is affected by environmental conditions. Children 6 to 10 years old from economically disadvantaged families have higher cortisol levels than advantaged children (Lupie et al., 2001), and adult women living under financial strain have unusually high levels of cortisol in the evening, while most adults have higher levels before noon (Grossi et al., 2001). Despite these findings, there is no simple relation between cortisol level and reaction to an aversive event or self-reported mood. When adults were administered either cortisol (20 or 40 milligrams) or a placebo and then asked to rate unpleasant and neutral words and pictures and to describe their mood, there was no relation between their mood or ratings of the stimuli and their blood levels of cortisol (Abercrombie et al., 2003).

Some temperamental categories may be derivatives of unique anatomical features. Adults with a larger than average volume of the right anterior cingulate, for example, reported more frequent bouts of worry and shyness with strangers (Pujol et al., 2002). The volume of the right medial ventral prefrontal cortex in monkeys, an area that is often active in anxious adults, is a heritable feature (Lyons et al., 2002).

It is also possible that the month of conception influences temperamental qualities through the secretion of melatonin in the pregnant mother, a molecule that can affect the fetus. If conception occurs at the end of summer (late July through September), the embryo's brain will be maturing as the hours of daylight are decreasing rapidly and the mother will be secreting more melatonin. Adults who were conceived during these late summer months have lower levels of serotonin, are at somewhat higher risk for schizophrenia (Chotai et al., 2000), and more likely to be shy in the preschool years (Gortmaker et al., 1997).

The immaturity of our current knowledge relating brain chemistry to human psychological states frustrates attempts to posit a lawful relation between a chemical profile and a particular temperament. Because genetic variation probably accounts for less than 10 percent of the variation in most human behaviors, it is unlikely that any single allele for a neurotransmitter level or receptor distribution will be the basis for a temperamental type.

Although no current temperamental bias can, at the moment, be defined by a specific neurobiological profile, the biology of the amygdala and its projections is likely to contribute to temperaments because this structure is influenced by many of these molecules. In addition, the amygdala has been implicated as critical in creating variation in the reaction to unfamiliarity.

The Amygdala

Despite its small size, the amygdala, less than 4 cubic centimeters in humans and relatively mature by the eighth prenatal month, consists of many neuronal collections, each with a distinct pattern of connectivity, neurochemistry, and functions. Each collection projects to at least 15 different sites and receives inputs from about the same number of regions, resulting in about 600 known amygdalar connections. Although a simplification, most anatomists conceptualize the amygdala as composed of three more basic areas: the basolateral, corticomedial, and central areas (Petrovich, Canteras, and Swanson, 2001; see Figure 3.1).

The *basolateral area* receives rich thalamic and cortical inputs from vision, hearing, smell, taste, and touch, and some input from the viscera. In addition, it is reciprocally connected to cortical sites, parahippocampal region, hippocampus, hypothalamus, basal ganglia, brain stem, and the bed nucleus of the stria terminalis. The behavioral reactions of flight or attack in response to a threat, and limb movement during states of arousal, are influenced by projections from the basolateral nucleus to the ventromedial striatum and ventral pallidum (Fudge et al., 2002).

The reciprocal connections between the basolateral nucleus and the perirhinal cortex might explain why unexpected or discrepant

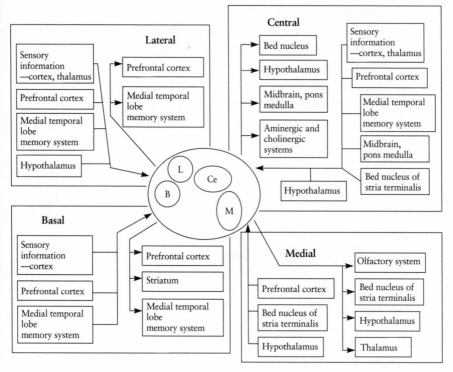

3.1 Schematic illustration of afferent and efferent connections of four
areas of the amygdala. (Adapted from Pitkanen, 2000.)

events are often remembered for a long time. The perirhinal cortex
sends information to the entorhinal area which, in turn, sends
information to the hippocampus to facilitate the storage of an experience (Kajiwara et al., 2003). Because the amygdala, which is
activated by discrepant events, primes the perirhinal cortex, the
probability of the latter evoking activity in deeper entorhinal neurons is enhanced; therefore, a discrepant event is more likely to be
registered in long-term memory (Seidenbecher et al., 2003).

The reciprocal connections between the prefrontal cortex and
the basolateral area permit the former to attenuate the level of
amygdalar arousal (Rosenkranz and Grace, 2002). It is possible,

therefore, that a chronically high level of anxiety could be due to compromised function of the prefrontal cortex. Because the prefrontal cortex underwent such a dramatic expansion in human evolution, older children are far better able than monkeys to control a strong emotional reaction to a threatening or unfamiliar event (Barton and Aggleton, 2000).

The *corticomedial nucleus* receives primarily olfactory and taste information and projects to the hippocampus, thalamus, hypothalamus, and the central nucleus.

The *central nucleus,* like the basolateral, receives some input from taste, vision, hearing, and the viscera and, most important, from the basolateral and corticomedial areas. The central nucleus is the origin of a large number of projections to the bed nucleus, cortex, basal forebrain, medial and lateral hypothalamus, brain stem, and autonomic nervous system. Unlike the projections from the basolateral area, which mediate gross motor reactions of fight or flight, projections from the central nucleus are more responsible for changes within the body, especially activation of the autonomic nervous system, the secretion of hormones, as well as alterations in muscle tone and distress cries.

The targets of the projections from the central nucleus mediate a variety of consequences. Activation of the bed nucleus leads to the secretion of stress hormones (for example, cortisol); activation of the parabrachial nucleus influences the parasympathetic nervous system; activation of the central gray leads to postural changes and distress vocalizations; activation of the lateral hypothalamus provokes the sympathetic nervous system; and activation of the reticularis pontis caudalis potentiates the startle reflex (Maren, 2001).

Some investigators suspect that the central nucleus is activated primarily by acute events to produce transient reactions, while more continuous threats activate the bed nucleus of the stria terminalis and result in a more chronic state of arousal. The level of CRH in the central nucleus is correlated with an animal's reactivity to a phasic event; level of CRH in the bed nucleus appears to be correlated with reactivity to a chronic stressor (McNally and Akil, 2002; Walker, Toufexis, and Davis, 2003).

The conditioning of potentiated startle or body immobility to a discrete conditioned stimulus (for example, a tone associated with shock) requires the central but not the bed nucleus. However, potentiation of the startle reflex in a rat first exposed to a bright light for 20 minutes requires the bed nucleus but not the central nucleus (McGaugh and Cahill, 2003; Brunzell and Kim, 2001). These two functions are analogous to the contrast between the state evoked by confronting a large snake on a forest trail and the state created by a beloved's prolonged illness. The central nucleus is needed to generate the biological changes defining an acute state provoked by a discrete event. The bed nucleus maintains a prolonged state of uncertainty to a chronic threat.

However, this rule, like most rules in biology, has exceptions. A rat with an intact bed nucleus but an inactivated central nucleus will freeze in response to the smell of fox feces without any prior conditioning. Apparently the freezing is a biologically prepared reaction to this olfactory stimulus (Fendt, Enders, and Apfelbach, 2003). It is important to appreciate, however, that neither the amydala nor bed nucleus acts as a whole, and one must specify which component participates in a particular reaction.

The neural activity that accompanies detection of an unfamiliar event is distinguishable from the activity that leads to subsequent biological or behavioral responses. The latter are the phenomena used to infer a state of fear, uncertainty, or surprise. The behavioral categories *inhibited* and *uninhibited* refer to the stable propensity to avoid or to approach unfamiliar events, and not to the recognition of their discrepant properties. The variation in behavior is more clearly a function of activity in the amygdala, rather than activity in the cortical areas that detect the fact that an event is discrepant from the long-term store of knowledge. These cortical areas include the parahippocampal region (combining perirhinal and entorhinal cortex and the parahippocampal gyrus) and prefrontal sites. These structures and the hippocampus are reciprocally connected to the amygdala. Thus, it is conceptually difficult to separate the differential contributions of the amygdala and the cortical structures to the detection of unfamiliarity, but easier to implicate the former in accounting for the avoidance or distress following the detection of unfamiliarity (Knight, 1996; Vinogradova, 2001; Rolls et al., 2003). All sighted children can perceive the discrepant features of an approaching adult wearing a mask. But because only some retreat or cry, it is reasonable to assume that those who show these reactions have a more excitable amygdala.

The dependent measures scientists select influence the brain sites they emphasize in their explanations. Consider a 2-year-old playing happily with toys who sees a person wearing a gorilla costume enter the room. The child initially startles, freezes, cries, experiences a rise in heart rate and feeling of tension, and flees to a parent. The scientist interested in explaining these reactions will emphasize the amygdala. But if the interest were in accounting for the perception of the discrepant quality of the intruder, the parahippocampal region would receive greater attention.

Valence

Neurobiologists are fond of writing that activity in the amygdala and its projections adds *emotional valence* to an event. Hence, it is useful to specify the biological referents for this psychological term. Activation of the parahippocampal region and basolateral nucleus provokes, through connections to the association cortex, components of a person's past history with an event, while excitation of the central nucleus leads to activation of motor, endocrine, and visceral targets. Feedback from the bodily activity to the orbitofrontal prefrontal cortex (OBPFC), combined with the historical associations, creates a brain state that, in humans, can pierce consciousness. The person's awareness of a change in bodily sensations motivates an attempt to interpret the unexpected feeling by comparing it with the immediately prior state and visceral schemata for the person's usual state. The outcome of these comparisons, instantiated in a new brain state, marks the event as salient and is the biological referent for the term *emotional valence* (Bindra, 1959).

But the emotional valence of an experience often depends on the particular comparison between what is occurring and what was anticipated. A noon temperature of 50 degrees while on a winter vacation in the Caribbean is experienced as unpleasant. The same temperature in Chicago on a windless mid-January day is felt as pleasant. Thus, a particular event need not have the same valence across situations. The brain state of a monkey who has received three raisins, for example, depends on whether it expected one raisin or six. A child who expects to be deprived of television because of an act of disobedience will regard a parental frown instead of the deprivation as having positive valence. The same parental frown will have negative valence if the child expected no consequences at all. Hence, with a few notable exceptions, states of pleasure and displeasure are not inherent properties of most events.

In one study, monkeys learned to associate a particular visual stimulus with the amount of juice they would receive—either low, moderate, or high—and to make a particular arm movement in order to receive the juice. One set of neurons in the striatum showed the greatest increase in activity when the visual stimulus signaled the delivery of the largest amount of juice, but different neurons became active when the stimulus indicated they would receive the least amount of juice. The authors concluded that striatal neurons were responding to the amount of anticipated reward. However, the neurons might be responding to a violation of expectation. One group of neurons were activated by a stimulus that signaled less juice than they had been receiving on two-thirds of the trials, while another set was activated by a stimulus informing them they would receive more juice than they received on two-thirds of the trials (Cromwell and Schultz, 2003; Miranda et al., 2000).

In a similar study, adults in an fMRI scanner who expected drops of sweet juice to be delivered following a motor response showed increased blood flow in the ventral striatum when the juice was not delivered because the discrepant experience primed the amygdala which, in turn, activated the striatum (Bevins et al., 2002; Berns et al., 2001).

Danger or Novelty?

The fundamental nature of the events most likely to activate the amygdala and the behavioral and psychological states that follow are a source of disagreement. The central issue is whether the amygdala reacts primarily to imminently dangerous events to produce a state of *fear,* or to unexpected or unfamiliar ones to produce a state of *surprise.* Potentially dangerous events should create distinctly different states than unfamiliar ones because not all unfamiliar events pose a threat, and some threats are neither unfamiliar nor unexpected. Residents of northern Bali distinguish between the extreme surprise caused by an unexpected event (called *tekajut,* which can occur in response to a sudden, loud noise, unwelcome news, or a violent change in posture, and means that the person might lose his soul and become ill) and a fear of physical harm (Wikan, 1989).

A female diana monkey issues a distinct vocalization to the unexpected alarm call of a male leopard. However, she does not vocalize to the same call a few minutes later, even though the leopard is still a threat, because now the call is expected. The same monkey would have vocalized if the second sound had been an eagle's shriek (Seyfarth and Cheney, 2003). The diana monkey vocalized when the potentially dangerous auditory stimulus was unexpected but not when the same sign of a dangerous event was anticipated.

Most vertebrates typically display a vigilant posture or a stereotyped motor reaction to particular events. For example, a lizard confronting another lizard with a dark spot on the skin near the eye shows an immediate rise in norepinephrine and dopamine and, in addition, assumes a passive posture (Korzan, Summers, and Summers, 2002). Rats, dogs, monkeys, and humans, on the other hand, have a stronger biological bias to react to events whose features are discrepant from their corpus of acquired knowledge. Mammals were the first animals to develop an elaborated set of amygdalar nuclei (Greenberg, Scott, and Crews, 1984). In an essay now over 70 years old, Valentine (1930) reminded readers that all mammals show distinct responses to the unfamiliar, and he cited Kohler's description of a chimpanzee's extreme behavioral reaction, interpreted as fear, in response to toy animals with black-button eyes (see Hebb, 1946; Bronson, 1970).

The original, and still cited, support for the claim that the amygdala is responsible for the behavioral response to discrepant events is the demonstration that monkeys with lesions of the amygdala and surrounding tissue eat unfamiliar foods that normal monkeys avoid (Kluver and Bucy, 1939), a discovery replicated with greater elegance recently (Amaral et al., 2003). Even though the lesioned animals approached unfamiliar objects and foods that were not dangerous, for more than 30 years scientists nonetheless have focused on the amygdala's participation in creating a fear state.

The roots of this view are in Brown, Kalish, and Farber's (1951) report, now over a half-century old, that the magnitude of a rat's body startle to an unexpected loud sound was enhanced when a light that had been previously paired with electric shock was presented just before the loud acoustic stimulus. These investigators assumed that rats should be "afraid" of a light that signaled an electric shock, and therefore the larger startles probably reflected a state of fear. However, the inference that rats are in a state of fear to a tone signaling shock is marred by the fact that it is not necessary to use electric shock as the unconditioned stimulus in Pavlovian conditioning of bodily freezing. Conditioned freezing can be established by pairing a conditioned stimulus with direct electric or chemical activation of the amygdala.

When the basolateral nucleus is activated by an electric shock applied to the animal's paws, it is intuitively reasonable to assume that the rat becomes fearful to the conditioned stimulus that preceded it. However, that intuition is less compelling when the basolateral nucleus is activated by chemicals, for this intrusion does not create the unpleasant sensations that accompany a shock to the paws. The function of the shock is to render the conditioned stimulus perceptually salient. As a result, the association between the conditioned and the unconditioned stimulus is established. It is disconcerting to learn that the amygdala of completely anesthetized rats was activated by a conditioned stimulus that had been paired with electric shock. It is conceptually troubling to call an unconscious animal fearful (Rosenkranz and Grace, 2002).

Pavlov (1927) did not call the dogs he conditioned to salivate to a tone "hungry" because the neural circuit activated by the conditioned stimulus that signaled food powder need not produce hunger. Similarly, a reflex smile in response to a friend's morning greeting need not be a sign of happiness, and a conditioned vaginal secretion to an erotic movie need not be a sign of sexual arousal. Had Brown, Kalish, and Farber measured the changes in cortical neurons produced by the conditioned stimulus, they would have found that it produced an alteration in the frequency of neuronal discharge. But the intuition that this brain reaction was a sign of fear would have been less obvious.

However, later work by LeDoux (1996), Davis (1994), and other scientists demonstrated that the thalamus and the lateral and central nuclei of the amygdala were needed in order to acquire a conditioned reaction of body immobility, potentiated startle, and increased heart rate to a neutral stimulus that had been paired with electric shock. This elegant research persuaded many scientists that the conditioned stimulus created a state of fear.

The anthropomorphic inference that a rat is "afraid" when it hears a tone that had signaled electric shock was attractive because the concept of fear played an important theoretical role in the first half of the twentieth century when psychoanalytic theory was popular. Fear seemed closely related to anxiety, and Freud had made anxiety the central culprit in the neuroses. Before Freud, a child who conformed to parental requests, was cautious in dangerous situations, and remained quiet with adult strangers was regarded as having a good character. After Freud, this child was classified as anxious.

Fear, along with hunger and sex, was regarded by behaviorists as an important drive state that catalyzed learning and new habits and could be manipulated in the animal laboratory. In addition, the Cold War threat of nuclear catastrophe, civil unrest during the Vietnam War, and warnings of air and water pollution, global warming, and, more recently, the attack on September 11, 2001, created a chronic uncertainty among many Americans. As a result, fear ascended to a position of prominence in the laboratory, and the stage was set for elegant analyses of the neural circuits that mediated this state.

Is Fear a Module in the Amygdala?

The accumulating evidence persuaded Ohman and Mineka (2001, 2003) that the amygdala reacts primarily to signs of potential harm rather than to unfamiliarity. They suggested that all animals inherit a fear module, located in the amygdala, that reacts, without conscious awareness or cognitive control, to events that pose a threat to the integrity of the body (snakes are presumed to be a classic example of a fear-evoking event). This hypothesis treats fear to particular events as a biological essence, analogous to positing an automatic hunger when blood glucose or lipid levels change, independent of the person's conscious thoughts or feelings.

There are serious problems with this argument. First, the behavioral reactions of most monkeys, chimpanzees, and human infants to a snake are not seriously different from their reactions to a host of unfamiliar events that are completely harmless (for example, a tortoise or dried seaweed). The British psychiatrist Isaac Marks (1987) described the terror his 2-year-old son displayed when he first saw thousands of dried skeins of seaweed. Fortunately, the boy lost his fear after repeated exposures to this novel scene.

Monkeys born and reared in a laboratory and therefore protected from contact with live snakes showed a longer period of motor inhibition to both a live and a toy snake than to blue masking tape, but only on the first testing session. By the second and third sessions the animals showed no more restraint to the snake than to the masking tape. Further, only 30 percent of the monkeys showed a more prolonged withdrawal to the live snake than to the tape, and the majority of animals failed to show any difference in duration of withdrawal (Nelson, Shelton, and Kalin, 2003). If snakes were a biologically potent incentive for a fear state, the motor restraint should not have habituated so quickly, and a majority of the monkeys—not 30 percent—should have shown a prolonged period of withdrawal.

School-age children from a Dakota Indian tribe in Manitoba, recalling the single most frightening event of their earlier years, most often named either a large domestic animal (a bull or horse) that had actually frightened them, or a ghost or witch-like figure they believed carried children away for disobedience—a threat Dakota parents used to socialize their children. Very few children named snakes, even though snakes are common in this area (Wallis, 1954; Means, 1936; Røskaft et al., 2003). Large groups of children from 6 different countries rated "loss of a parent" and "personal embarrassment or denigration," rather than snakes or spiders, as the experiences most likely to make them fearful (Yamamoto et al., 1996).

The replies of a large group of Norwegian adults—identical twin pairs, along with their spouses, siblings, and children—to questions about feared events revealed four relatively independent factors. The first two factors, which had most of the variance, referred to events that they had learned were causes of physical harm or death (a car accident, serious illness, being trapped in a boat in deep water). The feared events named by the third factor referred to social criticism for violating community standards. The feared targets in the fourth factor, which had the smallest amount of variance, included snakes, worms, rats, and spiders, and the environmental contribution to these fears was more substantial than the genetic contribution (Sundet et al., 2003).

Over 2,000 adults from six different world regions (northern Europe, southern Europe, North America, South Africa, Asia, and Africa), when asked to recall the circumstances in which they felt each of 6 unpleasant emotions (fear, disgust, sadness, shame, guilt, and anger), reported that a feeling of fear was most often provoked by events that were unfamiliar or unexpected (Scherer, 1997).

The experience of a friend of the authors who has a phobia of birds supports this argument. The woman dates the origin of her fear to an afternoon when, as a 7-year-old, she was watching Hitchcock's film *The Birds,* in which large flocks of birds attack humans. The woman remembers feeling very surprised by the fact that birds, which she had regarded as benevolent and beautiful, could be aggressive toward humans. This sharp disconfirmation of her childhood belief activated the amygdala, and the idea of harm became associated with the amygdalar activation and its somatic consequences. However, had she not been surprised by the birds' behavior, the phobia would not have developed.

The same mechanism explains why a 5-year-old boy developed a phobia of buttons. The boy went to the front of the room to retrieve buttons from a bowl in order to finish a teacher-assigned task. He slipped as he reached for the bowl, and all of the buttons spilled over him. This experience surprised and embarrassed him, and a phobia, which began that day, lasted for 4 years (Saavedra and Silverman, 2002).

Snakes are likely to become a feared target because of their discrepant features. They have an uncommon skin covering, an atypical ratio of head to body, and an unusual way of moving. The authors of the tree of knowledge allegory in the Old Testament probably chose the snake as tempter because of these uncommon physical properties. But discrepant stimuli elicit surprise, not fear, unless they cannot be assimilated. When surprise turns to fear following the unexpected encounter with a snake, it is because of symbolic associations. Snakes do not evoke any behavioral signs of fear in 6-month-old infants because they have not acquired the relevant knowledge; they have not yet learned the usual features of animals and do not know that some snakes are dangerous.

Adults from parts of south and east Africa hold a benevolent view of snakes. They believe each body contains a snake that protects it from disease and pollution (Green, 1996). Priests in ancient Egypt used artifacts in the shape of a snake to ward off disease, and surviving Minoan statues of goddesses with snakes coiled around their arms have a serene, not a fearful, expression. Historical and ethnographic records suggest that, over the course of human history, more adults have feared the spirits of their ancestors, witches, and sorcerers than snakes (Frazer, 1933, 1934, 1936).

A critical fact in the controversy over whether signs of danger or unfamiliarity preferentially excite the amygdala is that select neurons in the amygdala, bed nucleus, parahippocampal region, hippocampus, cortex, and brain stem respond reliably to unexpected or discrepant events *whether or not they are potentially harmful* (Habib et al., 2003; Wilson and Rolls, 1993), as well as to laughter, jokes, and sugar water (Wild et al., 2003; Zald et al., 2002b). The reactivity of these neurons habituates, often rapidly, as the event loses its novelty and its capacity to evoke surprise (La Bar et al., 1998; Ramnani et al., 2000).

In addition, the basolateral nucleus of the rat amygdala is activated by a conditioned stimulus that signals an unexpected opportunity for sexual behavior (Kippin, Cain, and Pfaus, 2003). And rats who see another animal receiving electric shock—an unusual experience for observer rats—show increased CRH in the central nucleus of the amygdala (Makino et al., 1999). The increase in delta power in the amygdala of squirrel monkeys was larger when the animal was in a situation in which unpredictable shock had occurred, compared to one in which a stimulus always signaled shock and therefore the aversive event was expected. The increase in delta power was even greater when the animal was seated beside another monkey—a novel event for the animal (Lloyd and Kling, 1991). Thus, delta activity in the amygdala was maximal when the situation was ambiguous, rather than when it was potentially harmful.

Adults playing a game against an experimenter showed activation of the amygdala during the brief period when they had made a choice that entailed risk of loss but did not yet know the outcome of their decision (Kahn et al., 2002). Playing roulette creates a state of excitement or uncertainty, not fear.

When rats were presented many times with a light followed by a tone which, in turn, was followed by food, they showed an orienting reaction when the light appeared without the tone because omission of the tone was discrepant from their past experience. Hence, these rats learned to associate the appearance of the light with the delivery of food. Rats without an amygdala failed to associate the light with the delivery of food because they were not alerted by the discrepancy of the light appearing without the tone and therefore did not orient to the light (Holland and Gallagher, 1999; Gallagher and Holland, 1994). Unfamiliar events can be powerful incentives. Infant rat pups who experienced a novel environment only 3 minutes a day over the first 21 days of life became adults who were behaviorally and physiologically different from controls (Tang and Zou, 2002).

Adults exposed to varying intensities of heat applied to the skin—from no heat to a very painful stimulus—showed equivalent amygdalar activation to the application of no heat and very intense heat because these events were the least expected in the series. If the amygdala were biologically prepared to react to an imminent aversive event, activity should only have been enhanced by the painful stimuli (Bornhovd et al., 2002).

Similarly, adults conditioned to expect electric shock to the tibial nerve when a flashing red light occurred did not show greater amygdalar activity to the red light than controls who experienced a random pairing of light and shock (Cheng et al., 2003). If the anticipation of pain following a conditioned stimulus elicits a fear state, as some presume, the experimental group should have shown increased amygdalar activity to the conditioned stimulus. However, if the amygdala reacts primarily to unexpected events, we might not expect any difference in amygdalar activity between the groups because the flashing red light meant that the subject would receive electric shock.

Adults in an fMRI scanner looking at familiar and unfamiliar faces with neutral expressions showed greater amygdalar activation to the new faces, even though no face presented a fearful, angry, or disgusting expression (Schwartz et al., 2003a; Roitman et al., 2001). Activity in the left amygdala was greater to angry faces if the direction of gaze was averted than if the angry face stared at the subject directly because the former is more ambiguous, even though the latter is more threatening (Adams et al., 2003).

One reason why masked presentations of faces with a fearful expression provoke amygdalar activity, and why fear faces usually provoke more activity than angry faces, is that the former expression is more ambiguous, not because it provokes a state of fear in the subject (Whalen, 1998). A similar process can explain why adults who had to decide on the gender of a photo (rather than the emotional valence of the facial expression) showed amygdalar activity to faces displaying happiness. Because the face had no cues from hair or dress, the detection of gender was very difficult (Keightley et al., 2003). All of this evidence supports the view that unfamiliar or ambiguous events have a greater potential to provoke the amygdala than potentially dangerous ones, as long as the cortical areas evaluating the degree of discrepancy are intact and can inform the amygdala of this fact.

One or Many Fear States?

The assumption that amygdalar activity mediates a single fear state is also open to challenge. First, animals without an amygdala displayed some behaviors scientists regard as signs of fear. For example, rats without an amygdala froze in response to the odor of a predator (Fendt, Enders, and Apfelbach, 2003; Wallace and Rosen, 2001) and defecated in a place where they had been shocked on an earlier occasion (Antoniadis and McDonald, 2000). Two-week-old monkeys without an amygdala screamed at other monkeys, although they showed no sign of fear to unfamiliar objects (Amaral et al., 2003). And guinea pigs with no amygdala, and only a brain stem and cerebellum, acquired a conditioned eyeblink response when electric shock to the orbital area was the unconditioned stimulus (Kotani, Kawahara, and Kirino, 2002). These facts suggest either that freezing and defecation in rats and screaming in infant monkeys do not reflect fear or, more likely, there are many fear states, each mediated by a distinct biology.

A second reason to doubt a single basic fear state comes from studies of two highly selected rat strains called Roman high- and low-avoidant. The former strain quickly learned to avoid electric shock. The low-avoidant rats, by contrast, did not learn this response quickly, partly because they froze when placed in the unfamiliar compartment where the shock was delivered. But the two strains did not differ in degree of exploration of an unfamiliar box or in time spent in the brightly lit arms of an elevated maze—two procedures assumed to measure a state of fear. Moreover, the low-avoidant rats, whom some might regard as less fearful because they did not learn to avoid shock easily, had more neurons in the central nucleus of the amygdala that stained for corticotropin-releasing factor and showed the largest potentiated startle when placed in a compartment where they had been shocked (Yilmazer-Hanke et al., 2001).

Study of two other selectively bred rat strains—called Maudsley high- and low-reactive—also invites rejection of a single fear state. The high-reactive (HR) strain was selected because these animals defecated when placed in an unfamiliar area. The low-reactive rats (LR) did not defecate in the same area. The original interpretation was that the HR animals were more "emotional" in an aversive environment than the LR rats, and their emotionality was accompanied by defecation. A quarter century of research on successive generations of the two strains has uncovered the error in that interpretation. The LR rats failed to defecate because they had much higher levels of norepinephrine in the heart, spleen, and colon due, presumably, to greater sympathetic innervation of these targets. Sympathetic innervation of the colon led to constriction of smooth muscle and, as a result, defecation was blocked.

Thus, the more accurate description of the LR strain is that they had higher levels of sympathetic tone in the colon, rather than that they were low in emotionality. Further, the LR animals had larger baseline startles in response to loud noises and spent less time contacting novel objects—responses that are inconsistent with a presumption of low emotionality (Blizard and Adams, 2002).

The Primacy of Unfamiliarity

A basic property of the basolateral area of the amygdala is a preparedness to receive information from the sensory cortex and parahippocampal region indicating that a current event deviates from the stimulus surround (a sound breaking the silence, for example), or from the agent's long-term store of representations (a picture of a human head on an animal's body), or from the context in which the event normally appears (a bar of soap on a dinner table). These three classes of events probably create different states (Dean, Redgrave, and Westby, 1989). *Alerted, surprised,* or *vigi-*

lant are possible names for the psychological states created by these experiences (Whalen, 1998). It is less clear what name or names are most appropriate for the underlying brain states. The main point is that the brain and the emergent psychological states that define these varied forms of surprise are different from those that are the foundation of varied forms of fear.

Rats and humans differ in their reactions to unexpected or unfamiliar events in two important ways. Only in humans do these events generate a change in conscious feeling tone that is interpreted in light of the quality of the feeling, their store of knowledge, and the immediate context. Second, the expanded prefrontal cortex of humans permits more effective control of the amygdala and its projections. Neuroscientists name the state of amygdalar activation *fear*. Most psychiatrists name the interpreted change in feeling tone *fear*. Each group is entitled to naming rights, but these two uses of the term are not synonymous.

The suggestion that unfamiliar events are the primary incentives for amygdalar activity provides a more coherent explanation of the evidence than the popular assumption that this structure is biologically prepared to create a fear state to imminently dangerous events. The Kluver-Bucy monkeys approach unfamiliar foods because of the absence of amygdalar projections to targets that produce motor restraint to novelty. Rats with no amygdala fail to learn that a light signals food because they are not alerted by omission of the tone that normally followed the light. Most of the fMRI or PET data in humans can be explained by assuming that the amygdalar activation was caused by unfamiliar or unexpected events.

This hypothesis can be applied to the Pavlovian conditioning of rats as well. The unexpected tone that is the conditioned stimulus activates the amygdala, which, in turn, creates an alerted state. The unexpected electric shock moments later enhances amygdalar activity, and as a result neuronal representations of the tone and shock become associated. The later presentation of the tone evokes the association between tone and shock and provokes the central nucleus to send projections to the central gray to produce immobility.

Neuroscientists use the word *fear* as a conceptual invention to explain the lawful relations among the tone, shock, and freezing; it may not describe the animal's psychological state. The decision to use this word to name the cascade of events that begins with a tone signaling shock and ends a second later with an immobile rat is based on the premise that an animal's brain state during this brief interval resembles that of a person who has been told he has a malignant cancer. We suspect this premise is flawed and suggest that neuroscientists should have chosen a term with semantic features more closely related to *surprise,* because the tone, as well as the electric shock, were both unexpected events that activated the amygdala and related structures. A brief, unexpected loud sound while eating breakfast would cause a startle, a moment of immobility, and a small change in heart rate, but most individuals would say that they were surprised, not afraid. If the sound occurred several times over the next hour, the bodily reactions would cease to occur because the person was no longer surprised. Recall the female diana monkey and the leopard call.

A person held hostage for 6 months by terrorists might feel continually fearful but probably not continually surprised. We believe that the primary function of the amygdala is to initiate a cascade of physiological reactions to unfamiliar or unexpected events. The degree of amydgalar activation is correlated primarily with the magnitude of discrepancy or degree of unexpectedness of an event, rather than with its degree of danger or level of aversiveness (Cahill and McGaugh, 1990; Kapp, Supple, and Whalen, 1994). Although Davidson (2003a) suggests that most unfamiliar events have an aversive valence, it is not obvious to us that unexpected encounters with a new food, friend, movie, place, or news story are less frequent each day than unexpected events that are potentially harmful.

The amygdala–fear connection survives in part because of the commendable desire of investigators to find a material foundation for the psychiatric syndromes called the anxiety disorders (see below). The elegant experiments with rats, which pinpoint the contribution of the amygdala to conditioned freezing, provide a material basis for the vaguer concept of human anxiety. When scientific evidence replaced religious writings and sentiment as the foundation for ethical norms, medical practices, and legislation, citizens felt more secure if the rationale for societal practices rested ultimately on what scientists had learned about nature. It is not surprising, therefore, that individuals experiencing chronically high levels of worry over their children, health, employment, social status, and ethical integrity would feel a bit better when they learned that their worry could be traced to their amygdala and was not a vague, ethereal construction of their mind. The assumption that anxiety is a stable, material state of brain/mind renders individuals receptive to taking medicine as therapy. But if one conceives of anxiety as a dynamic state that can be altered in diverse ways, then exercise might be chosen. And aerobic exercise does reduce the intensity of worry over unexpected bodily sensations that might be interpreted as a sign of anxiety or mental illness.

Readers familiar with the history of psychology will appreciate that a state of surprise had been regarded as a basic emotion and that events that provoked this state could function as rewards (Kaplan, Fox, and Huckeby, 1992). Daniel Berlyne, who explored this idea in the 1960s, suggested that both animals and humans were attracted to novelty because of a curiosity drive (Charlesworth, 1969). Neuroscientists forgot this history when they began their studies of conditioned freezing and potentiated startle because they wished to understand fear, not surprise. They assumed that rats were capable of fear because of the hope that the Pavlovian paradigm would be a useful model for human anxiety disorders.

But the primary origins of anxiety in humans are thoughts that imply the individual might, sometime in the future, experience bodily harm (including contamination), loss of property or an important relationship, or a threat to his conception of himself as a competent, honest, kind person entitled to dignity. Thoughts rich with symbolism, rather than conditioned stimuli for pain, are the most common basis for activation of the amygdala and its projections in humans. College students from 5 different countries (Germany, Mexico, Poland, Russia, and the United States) nominated black as the one color of 12 that was most symbolic of fear (Hupka et al., 1997), perhaps because one feels less certain in darkness.

Many individuals can acquire an avoidant reaction to objects or situations simply by being told that these events are possible threats. This knowledge can be learned without an amygdala. Over 80 percent of a sample of American college-age males born in the first decade of the last century reported a fear of masturbation because they had been told by their parents or others that this practice could harm the mind and body (Pullias, 1937). A fear of shrinkage of the sexual organs—called *Koro*—reached epidemic proportions in Guandong, China, in the mid-1980s (Tseng et al., 1988).

The amygdala is needed for the somatic components of an emotion—a rise in heart rate, blood pressure, muscle tension, respiration rate, secretion of cortisol, or sweating as well as flight or freezing to unfamiliar events. But humans interpret these unexpected changes in body sensations as meaning that they might be afraid, even when there is no immediate or anticipated threat. Thus, the states we call fear and anxiety in humans are seriously different from the animal states that are given the same name.

Over a half century ago, in a presidential address to the division of experimental psychology of the *American Psychological Association,* Frank Beach (1950) offered a light-hearted critique of his colleagues in a paper playfully called "The Snark Was a Boojum." Beach's point was that many psychologists interested in how humans learn new habits assumed, a priori, that the principles that explained how a white rat learned to find the goal box in a maze would turn out to be essentially similar to the principles that explained how a child learned to speak, read, and play soccer. That premise proved to be too ambitious. There is the possibility that a rat freezing to a tone that signals shock is an equally inappropriate model for many, but perhaps not all, forms of human anxiety.

Human Fear and Anxiety?

Agreement on the meanings of *fear* and *anxiety,* like the meanings of all words naming human emotions, remains elusive. We believe that the two basic components of human emotions are a brain state and an accompanying set of cognitive representations (even though some scientists attribute an emotion to the brain state alone). Three thorny issues center on (1) the class of events that induce varied brain-cognitive coherences; (2) differences between the state of anticipation and the state during, and immediately following, an event; and (3) differences between acute and chronic states.

The traditional distinction between fear and anxiety was based on whether the unwanted event was imminent or anticipated. English speakers use words like *afraid, frightened,* and *scared* to name the state generated by an imminent threat but prefer the words *worried, concerned,* and *troubled* to describe an anticipated threat. At least four psychological states are defined by a class of imminent event and the representations and somatic sensations it evokes:

- *Uncertainty over harm.* This state is created by events that are immediate threats to the integrity of the body. Among Americans, these include imminent physical attack, contamination, illness, hunger, accident, natural catastrophe, or panic attack.
- *Uncertainty over one's virtue.* This state is created by events that are immediate threats to the individual's conception of

self and include domination, rejection, criticism, task failure, loss of status, and violations of the individual's ethical standards. Plato suggested in the *Dialogues* that humans are capable of two kinds of fears: harm and a critical evaluation by others. However, Plato did not mention the following two classes of unwanted events.

- *Cognitive uncertainty.* This state is created by unexpected or unfamiliar events that are not understood immediately. These events create uncertainty over the origin and consequences of an event, or the actions the person should implement. The uncertainty generated by an inability to understand a discrepant event is different from the state created when one is uncertain over what response to make in a familiar situation. That is, the state created by an unfamiliar noise at midnight is different from the state that accompanies indecision over what to do when a friend has been insulting. Infants show no signs of fear when they can control the sudden movement of a jack-in-the-box, but cry when they cannot do so or when they do not know when the event will occur.

- *Conditioned uncertainty.* A fourth state is created by events that, through experience, have become conditioned signals for any one of the above three states. For example, if a person became faint while watching a violent scene on a movie screen, scenes with violence could become a conditioned signal for a fainting reaction through standard Pavlovian mechanisms.

A parallel quartet of states can result from anticipation of each of the above classes. The state created by a robber with a gun demanding your money or your life is unlike the state created by anticipating the same threat while sitting by a fire at home. The state produced by sharp criticism from a teacher for poor test performance is dissimilar from the state generated by anticipating the same disapproval. The state of a panic patient who unexpectedly experiences a rise in heart rate and blood pressure is different from the state of the agoraphoric who remains home because she anticipates a possible panic attack. Finally, the state created by seeing a violent scene on a movie screen that has become a conditioned signal for feeling faint differs from the state produced by anticipation of seeing the violent scenes while walking to the theater (Funayama et al., 2001).

Thus, there are at least eight different members of a family of related states we call uncertainty:

1. Uncertainty over imminent harm
2. Uncertainty over possible harm in the future
3. Uncertainty over a compromise in one's virtue
4. Uncertainty over a potential compromise in one's virtue
5. Uncertainty over a discrepant event present in the perceptual field
6. Uncertainty over encountering a discrepant event in the future
7. Uncertainty created by a conditioned stimulus for any of the above six states
8. Uncertainty created by possible encounter with a conditioned stimulus for the above states

Most patients with an anxiety disorder report two distinctly different sources of their state: their present-life conditions and worry over possible loss of control of one's emotions or the ability to deal with threat. The first refers to an imminent event; the second to an anticipated one (Diefenbach et al., 2001). Scientists who study conditioned freezing in animals are probing a component of state 7; most patients with social phobia are in state 4; states 2, 4, 6, and 8, which involve anticipations, are more common in humans than in animals.

The diary of Pitirim Sorokin illustrates the difference between a state of uncertainty regarding possible harm and the state created by certain harm. Sorokin, who was arrested by the Bolsheviks in 1918 as a counter-revolutionary, was told he would be executed within the next few weeks. Sorokin was certain that the threat would be implemented, for each night a group of prisoners were taken from their cells and executed the next morning. Sorokin recalled feeling anger, not fear, the evening he was informed he would be killed the next morning. "Do I fear this suffering? Not at all. Why, then, does all my organism, all my soul, all my self protest against this? Why do I feel so upset? Why? Not because I fear, but because I want to live" (Sorokin, 1924). Sorokin felt no fear when he was certain he would die; but weeks earlier, when he did not know whether he would be arrested, his consciousness was invaded with fear. The important point is that the state that accompanies uncertainty over possible harm is different from one that accompanies certain harm. There is honest controversy over the words we should use to name these two states, but scientists should distinguish between them. The ancient Romans had a choice of three different Latin verbs to describe the act of kissing, depending on whether the target of the affectionate gesture was a spouse, lover, or friend.

The human states we call fear or anxiety are interpretations imposed on a combination of thoughts of unwanted events and detected bodily sensations. If one is eating dinner at home and thinks of being hurt in a plane crash, somatic sensations could be elicited, but usually they are not. Further, on some occasions an individual might interpret the bodily sensations as meaning that they are anxious or fearful, even if no thought of an unwanted event occurred. That is, sometimes a thought elicits bodily sensations; at other times the sensations occur first and invite the interpretation that one is anxious. We believe that a temperamental bias is more likely for the second of these scenarios; life conditions and past history are more likely causes of the first sequence.

The English words *fear* and *anxiety*, like terms for most human emotions, were invented centuries ago to name a person's interpretation of a family of feelings associated with thoughts of an unwanted experience. It is unlikely that these words, which fail to contextualize their origins or targets, are more useful as scientific concepts than a host of other words in the public vocabulary, such as *smart, arrogant, stubborn, crazy,* or *debauched.* Scientists need to invent new terms that describe the separate brain and psychological states created by a particular class of events. For example, we need a concept to represent the brain state created by activation of the thalamus, amygdala, and central gray when a rat freezes in response to a tone that had been associated with electric shock. This state might be called the *conditioned central gray circuit,* or simply CCG. We should not use *fear* to name this brain state because this circuit can be activated by events that are not signs of imminent discomfort, like the unexpected appearance of a friend one has not seen for years.

Before there were neurobiologists, the semantic features of the word *fear* were a conscious state that combined awareness of an imminent threat to one's safety with felt somatic sensations. The discoveries of the last three decades have created a new semantic network for fear that includes rats, immobility, startles, and the amygdala as central nodes, but not conscious worry over criticism, loss, or failure. The semantic network for a word that names a feature, function, or property of an entity is linked to that entity; hence, its meaning can change if the word describes events from a different category. For example, the words *true* and *false* are only appropriate adjectives for statements, not for animals, people, or foods. The term *alienated* describes people; *stratified* applies to societies. Most speakers respect the linguistic rules of their community, although they understand that poets and novelists enjoy the liberty of applying a word normally linked to members of one class to those from a different category.

However, scientists, who do not have this freedom, occasionally use words like *fearful, sociable,* and *aggressive* to describe mice and rats, as if these words had the same meaning when applied to humans. It is even possible that use of the term *anxious* or *fearful* in sentences that describe a mouse who avoids a brightly lit area comes close to being a metaphor. When Dylan Thomas awarded the quality of *dominion* to death, readers understood that the poet's intention was to evoke an aesthetic response. But behavioral biologists who write that mice are aggressive or anxious would like to believe they are describing the animals' psychological state.

Anxiety Disorders

Although everyone has the capacity to anticipate an unwanted event, the intensity of the accompanying feelings is muted most of the time and does not interfere with the performance of the day's tasks. However, a small proportion who experience these feelings with greater intensity or regularity have a compromised ability to deal with daily responsibilities. These individuals are classified as possessing an anxiety disorder. It is important to appreciate that the current approved list of anxiety disorders reflects the consensual opinion of American and European psychiatrists born in the twentieth century. These categories are based on the verbal reports of individuals seeking medical help and are not derived from extensive research on random samples of adults from many cultures that include behavioral and physiological evidence. A blue-ribbon committee of physicians charged at the beginning of the eighteenth century with listing the major physical diseases would not have named cancer, diabetes, and a sclerotic coronary artery as primary diseases because these categories were not diagnosable from patient reports of symptoms. Equally important, the current set of anxiety disorders often ignores the social context in which the patient lives. This position implies that report of a chronic state of worry in an unmarried woman with four children, a marginal income, and residence in a crime-infested neighborhood is not fundamentally different from the report of the same degree of worry in a married woman with one child, a secure income, and a professional career.

Adults who failed to graduate from high school and who, as a consequence, usually have lower incomes and live in residential areas with higher crime rates have realistic bases for worrying about their health, status, finances, and security. It is not surprising, therefore, that anxiety disorders are more prevalent among economically disadvantaged groups than among those with financial security (Gallo and Matthews, 2003). The one-month prevalence of any of the anxiety disorders increases from 4.6 percent for adults belonging to the highest social class category to 10.5 percent for those in the lowest category—a ratio of 2 to 1 (Martin, 2003).

The modest relation between the probability of a diagnosis of an anxiety disorder and social class is not a new phenomenon. A group of 1,000 women from working-class families interviewed in the 1930s reported more fears than women from middle-class homes (Means, 1936). However, it is true that some anxiety disorders have a heritable biological component; for example, estimates of the heritability of panic disorder range between 0.5 and 0.6 (Merikangas and Risch, 2003). Contemporary Americans and Europeans report that the following situations make them anxious (Cox et al., 2003):

- Events that can harm the body (encounters with poisonous snakes, spiders, dirt, flying, infectious diseases)
- Situations where they might be judged in an unfavorable way by others (speaking in public or being observed by a stranger in public places)
- Loss or separation from a person who is a source of support or affection
- Unexpected changes in bodily feelings that are experienced as unpleasant

- Images or thoughts of a past traumatic event
- Irresistible urges to perform certain actions

Adults from other cultures, or historical eras, would expand this list to include possession by the devil, God's wrath, the actions of a sorcerer or witch, arbitrary imprisonment, bandits, plague, chronic hunger, forced conscription, torture, stoning by a crowd, or being sold as a slave.

The ancient Greeks, as well as Europeans during the Middle Ages, did not regard anxiety as a mental illness because worry over Zeus's actions, God's wrath, or social criticism for violating community mores were utilitarian emotions that guaranteed civility and obedience to local rules. Anxiety was an aid to adaptation rather than an alien emotion to be exorcised. Further, most individuals in premodern times lived with extended families who provided social support when they became worried. Single adults with few friends living alone in apartments in large cities represent a historically unique social condition that makes chronic anxiety a more likely phenomenon.

More important, the industrialization and geographic mobility that expanded during the late nineteenth century required a more individualistic, entrepreneurial spirit; hence, anxiety became a serious obstacle to self-actualization and an impediment to conquer. Although nineteenth-century Americans equated "good character" with the ability to control feelings of fear, Freud's suggestion that repression of anxiety led to serious symptoms transformed the halo of virtue worn by the courageous into a dark sign of mental illness. Plato would have been surprised to learn that contemporary Americans regard anxiety as a state that compromises adaptation and one that should be eliminated from human experience. As history replaced the scripts citizens had to follow in order to attain symbols of virtue, anger and anxiety exchanged positions in the hierarchy of unwanted feelings. Many contemporary Europeans and North Americans have decided that the primary goal of life is a state of "happiness" gained through satisfaction of appetite, actualization of talent, upward social mobility, a love object, family, and freedom from worry. Hence, the recognition that one is anxious is often interpreted as a sign of personal compromise that exacerbates, rather than mutes, a feeling of distress. St. Augustine would not understand the changes that history has wrought.

It is also possible that the historical events of the past two millennia have imposed a greater burden on modern citizens than they did on the ancient Athenians. Anger feeds on the past—through an insult, attack, or frustration imposed by another person or group —and it is easier to avoid situations that might generate anger. Anxiety feeds on the future—on the anticipation of events that might compromise a person's psychological and physical integrity. Because the number of unwanted events is so large, and the human mind so creative, it is easy to feel anxious. Finally, whether an experience evokes anxiety or anger often depends on the person's interpretations of the event. A less than civil greeting from a friend could evoke anger if a person decided that the friend intended to be insulting, but anxiety if the person interpreted the snub as a response to an earlier social error performed unknowingly. Because humans are biased to assume that they committed a moral transgression, anxiety is more likely in ambiguous situations where either anxiety or anger is possible. Although the Ojibwa Indians believed that illness was due to the acts of a sorcerer, those who became chronically ill felt anxious, not anger at the sorcerer, because they assumed that one of their ethical lapses provoked the malevolent action.

Because a majority of patients with an anxiety disorder have never experienced the events they fear—or, if they have, the frequency of encounter was low—conditioning theory would predict that their anxious state should have extinguished over time. Because it does not, we must ask why. There are several possible explanations.

When psychoanalytic and psychodynamic theories were popular a half century ago, many assumed that some life histories could create a chronic feeling of guilt over violations of one or more salient moral standards. It was presumed that one consequence of this guilt was an expectation of harm or loss as a symbolic punishment for the ethical lapse. Because the guilt was chronic, the anxiety generated by the possibility of punishment persisted. This explanation of anxiety is less attractive today, as the intensity of guilt over sexual and aggressive behavior has decreased but with no obvious decrease in the prevalence of anxiety disorders.

A second mechanism, which relies on the processes of classical conditioning, assumes that the individual experienced intense distress during one or more encounters with the feared event, and its mental representation persisted. Some human anxiety disorders are probably acquired through this Pavlovian mechanism, if we let a class of thoughts replace a tone or light as the conditioned stimulus and a salient increase in unpleasant feeling tone replace electric shock as the unconditioned stimulus. The woman with the phobia of birds experienced a sharp change in feeling tone when her belief that birds were harmless was disconfirmed. The conditioned stimulus that maintained the phobia was her rich set of cognitive representations for birds. Because her limbic reaction was intense, it took only a single trial to establish a Pavlovian association between thoughts of birds and a limbic reaction, which she interpreted as fear.

Many thoughts can become conditioned stimuli for a biological reaction, including abandonment, danger, task failure, violation of a moral standard, peer teasing, rejection, criticism, a disadvantaged status, lack of money, or insufficient time to accomplish a required task. Because the brain reaction and accompanying changes in feeling tone are necessary to establish the association between the thought and the interpretation that one is afraid, variation in susceptibility to activity in particular circuits should predict a vulnerability to one or more of the anxiety disorders.

A person's conscious feeling tone is a critical component of the state of anxiety, and activity in the orbitofrontal prefrontal cortex (OBPFC) makes an important contribution to this feeling. A large section of this area receives sensory information from olfactory, gustatory, viscera, somatosensory, and visual sources related to eating behavior, but this area is responsive to other events as well. For example, a smiling, attractive face evokes more activity in the OBPFC than a non-smiling, unattractive face (O'Doherty et al., 2003). The OBPFC also receives input from the insula and amygdala, especially the basolateral area, and sends a synthesis of this information to a part of the OBPFC called the ventromedial prefrontal cortex. This area has few sensory inputs, but is the origin of projections to the hypothalamus and brain stem (Price, 1999). This anatomy implies that frequent or intense changes in somatic sensations can be due to an excitable OBPFC or excitability in the sites projecting to or from the OBPFC (Zald, Mattson, and Pardo, 2002).

Vulnerability to Bodily Sensations

Each visceral target—heart, muscle, gut, labyrinth—sends its afferent impulses to the brain in distinct tracts to sites in the brain stem, amygdala, and the OBPFC, and each is influenced by distinct chemistries. It is reasonable to assume, therefore, that genetic differences among individuals can render a particular visceral target more or less reactive. Some individuals will have a more reactive heart; others a more reactive gut.

Two complementary mechanisms might explain a temperamental vulnerability to the collections of symptoms called anxiety disorders. The first assumes a single temperamental bias for brain arousal that is combined with each person's unique history to determine the particular symptom that develops. The most parsimonious form of this argument assumes that some children inherit a neurochemistry that lowers the threshold of excitability of a brain circuit—probably involving the amygdala and the OBPFC as components—that, when activated, leads to a change in bodily sensations and a conscious feeling interpreted as worry, fear, or vigilance. The interpretation the person imposes on this state, and the symptoms that appear, will be a function of life history. A child who had been consistently criticized for violating norms of cleanliness might develop a washing ritual; a child regularly criticized for task incompetence might become a social phobic; and a child who experienced a serious loss or frequent separation from a caretaker might develop separation anxiety disorder.

This hypothesis is appealing because of its simplicity, but it is seriously inconsistent with the fact that patients with compulsive symptoms have a different biology from those with social phobia, blood phobia, or snake phobia. Although the absolute risk is low, high-reactive infants are more likely than others to develop symptoms of social phobia during adulthood, but they are not more vulnerable to animal or blood phobias. Hence, a more reasonable argument is that distinctly different brain profiles underlie the separate anxiety disorders. This view asserts that individuals vary in susceptibility to distinctive brain and body states to particular events. The target that is feared is determined by the particular susceptibility.

For example, the sight of blood accompanying serious injury can evoke a vasovagal reaction, characterized by a sudden drop in blood pressure and a feeling that one is about to faint (Page, 2003). Therefore, individuals who inherit a susceptibility to a vasovagal reaction would be more vulnerable than others to acquiring a blood phobia but would not be particularly vulnerable to acquiring a phobia of snakes or public speaking. Three children in our study who had blood phobias had high vagal tone and no other fears. By contrast, individuals with a susceptibility to a brisk sympathetic reaction, accompanied by a sharp rise in heart rate and blood pressure, might be especially vulnerable to developing panic disorder. A person susceptible to extreme facial flushing, which often occurs when one anticipates a critical evaluation by others, might be especially vulnerable to developing social anxiety (Crozier and Russell, 1992; Gerlach, Wilhelm, and Roth, 2003; Mulken et al., 2001). Adults with a susceptibility to compromised vestibular function might be vulnerable to frequent bouts of dizziness and therefore to the development of PTSD (Ozer et al., 2003). However, each person's past history and cultural setting are always modulating factors.

Caucasian Americans and Europeans who feel anxious focus on their racing heart or feeling of suffocation, due partly to sympathetic lability. Asians more often report a feeling of dizziness, due in part to withdrawal of sympathetic tone (Hinton and Hinton, 2002; Park and Hinton, 2002). Each conscious feeling is inserted into an acquired belief network. North Americans and Europeans have learned that a racing heart might signal an imminent heart attack, a mental illness, or failure to control one's emotions; Mayan Indians living in northwest Guatemala believe that this sensation means that a life spirit has left their body.

Biological vulnerabilities are not restricted to unpleasant feelings. It is possible that a temperamental bias can motivate individuals to seek very particular pleasures. Children with a "sweet tooth" may actually extract more sensory pleasure from chocolate than others. The pair-bonding of prairie voles, with which we began this chapter, might be due to the fact that mating is inherently more pleasurable for them than for montane voles because of their unique distribution of receptors for vasopressin and oxytocin. A small number of adolescents pursue a particular career because they had an intense emotional experience surrounding a particular activity. Some leading cosmologists have reported that, as youths, they had intense feelings of awe when staring at the night sky. One adolescent in our sample described a powerful feeling of joy when she performed modern dance, and she has decided to become a professional dancer. Thus, in addition to the obvious contribution of life history, it is credible to suppose that a young person with a temperamental susceptibility to a particular feeling will associate that feeling with his interpretation of its source and, in so doing, select a vocation to pursue.

The anticipation of an infrequent or unexpected desired event is accompanied by a release of dopamine, and animal strains differ in dopamine metabolism and in the density and distribution of varied dopamine receptors when they unexpectedly encounter a new food or place (Comings and Blum, 2000). Thus, individuals with a temperament that leads to a greater than normal release of dopamine to an unexpected but desired event (or a unique distribution of one of the dopamine receptors) might actually feel more pleasure than others when they anticipate visiting a new place, adopting a new hobby, or establishing a new friendship.

Female mice, rats, and humans, compared with males, possess greater concentrations of dopamine in the striatum and a greater receptor density for dopamine in frontal, temporal, and cingulate cortex, due in part to the fact that estrogen potentiates dopaminergic function (Laasko et al., 2002; Mozley et al., 2001; Kaasinen et al., 2001). This fact invites an interesting speculation. If most males have lower brain dopaminergic activity than most females, and if unfamiliar events typically produce an increase in dopamine and a brief pulse of pleasure, more males than females might experience a special delight from novel experiences because their level of dopamine activity was lower initially. Women seem to derive greater pleasure than men from long-standing familiar friendships and somewhat less pleasure than men from completely new experiences. In addition to the obvious influences of socialization, biology may make a small contribution to this gender-based difference.

However, most people have a sparse and relatively inexact vocabulary for the subtleties of their feeling tones, but a rich vocabulary for the events that produce these feelings. Hence, they tell themselves and their acquaintances that what they desire is money, friends, travel, love, success, or higher status, and what they fear is flying, public speaking, strangers, or snakes. But these easily named events are only the conditioned incentives for the feelings that are the fundamental events the person seeks or avoids.

Summary

This chapter has presented a detailed discussion of the amygdala because we believe that the variation in motor activity and crying in response to unfamiliar sights, sounds, and smells that we observed among 4-month-old infants is due to excitability in this structure and its projections. However, as we noted earlier, the stimuli we presented recruited activity in other structures—for example, the parahippocampal region, which is reciprocally connected to the amygdala. Thus, it is possible that an important cause of the variation in motor activity and distress was differential excitability of the parahippocampal region. Hence, we do not suggest that the infant biases that lead to inhibited or uninhibited behavior are "located" in the amygdala, only that this structure makes a relevant contribution because it is the immediate origin of the behaviors that define high and low reactivity.

Now that the intellectual scaffolding for our work is in place, it is time to summarize what we have learned from the assessments of the 11-year-olds in our study.

4

BEHAVIORAL AND BIOLOGICAL
ASSESSMENTS

Our sample consisted of 237 Caucasian middle-class children born at term to mothers who experienced a normal pregnancy and delivery. All children had been assigned to a temperamental category at 4 months based on the protocol described in Chapter 1 and were observed for inhibited and uninhibited behavior when they were 14 and 21 months old. This group contained 92 children classified at 4 months as low-reactive (39 percent of the sample; 46 girls and 46 boys), 68 high-reactives (29 percent; 31 girls and 37 boys), and 77 children who belonged either to the low-motor–high-cry or high-motor–low-cry infant groups (32 percent; 42 girls and 35 boys). Their average age was 10.6 years (ranging from 9.8 to 12.4 years). Two-thirds were between 10.3 and 10.6 years, 5 children were 12 years old, and only 1 child was a little under 10 years old.

The primary assessment, about 3.5 hours long, was a laboratory session conducted in a room without windows by a young Caucasian woman who was unfamiliar to the children and did not know any of their prior history. These preadolescents knew they had visited the laboratory many times in the past and were aware that their parents hoped they would be cooperative. Therefore,

Table 4.1. Variables and sample sizes.

Order of collection	Sample size
Rating of inhibited behavior*	235
Spontaneous comments	235
Number of smiles	235
Resting heart rate	219
Blood pressure	234
Spectral analysis of heart rate	185
Ear temperature	234
Heart rate during test of auditory acuity	227
Wave 5	187
EEG power and asymmetry	228
Event related potential	205
Startle to the puff procedure	204
Corrugator to the puff procedure	212
Startle to the pictures	158
Corrugator to the pictures	183
Triads	224
Finger temperature	208
Height	234
Weight	234
Mother's Q-sort	222
Child's Q-sort	215

*Videotapes of two children were defective and could not be coded.

these boys and girls were probably more motivated to conform to the procedures than the average 11-year-old in a cross-sectional investigation.

Initially, the child and parent, usually the mother, came to the laboratory and met the examiner. The mother read a description of all the procedures to be administered, and the examiner explained them to the child. Thus, every child knew what was going to happen over the next few hours (Table 4.1). The parent then left the room and was given the choice of going to an adjoining room to watch her child over a monitor or to leave the building and return later when the battery was complete. A majority chose the latter option. Almost all children completed the battery, and very

Table 4.2a. Child's Q-sort (possible responses: "very much like me; a little like me; not like me; definitely not like me").

I feel really bad if one of my parents says I did something wrong.
I wonder a lot about what other kids think of me.
I worry too much about making mistakes.
Most of the time, I'm happy.
I like going to new places.
I get a little scared in new places.
I like to play with many children at recess.
I like raising my hand in class to answer a question.
If I want to see a friend, I will call on the phone.
I rarely have nightmares.
I don't like to read in front of the class.
I am a shy person.
I like to do risky things.
I like going on roller coasters at an amusement park.
I am a little nervous when I sleep over at a friend's house.
I like to play with 1 or 2 kids during recess.
If I want to see a friend, I like my parent to call.
I like to be first in gym class to try playing a new game.
I like to watch other kids try a new game before I try playing it.
If I have a question in a store, I like to have a parent ask the sales person.

few protested or complained about any discomfort. Some of the measures have less than the full sample because of technical failures or, in a few cases, refusal of a procedure. Some children could not control their breathing, which was required for the spectral analysis of heart rate, and these children were excluded from that analysis.

In addition to a laboratory session, the mother and child were visited at home by Christina Hardway, who asked them separately to rank different sets of descriptive statements, from most to least characteristic of the child. The mother's set of descriptors contained 28 items; the child's set contained 20 items (Tables 4.2a,b). Ninety-two percent of mothers and children filled out the Q-sorts, and the majority did so in their home under the guidance of Hardway, who had no knowledge of the child's temperamental category

Table 4.2b. Parent's Q-sort (possible responses: "very much like my child; a little like my child; not like my child; least like my child").

Does well in school.
Well-behaved with adults
Makes friends easily.
Handles responsibility well.
Outgoing with other boys and girls.
Doesn't like to fail at a task.
Refuses requests and rules at home.
Has a lot of friends.
Prefers one or two close friends to many friends.
Is a leader with peers.
Talks a lot with other children.
Loses temper when doesn't get own way.
Is sensitive to punishment or being reprimanded.
Shy with other children he/she doesn't know.
Has a lot of energy.
Easy-going.
Acts before thinking; impulsive.
Likes to try new things.
Concerned about what others think of him/her.
Becomes quiet, subdued in unfamiliar places.
Worries about what might happen in the future.
Gets angry easily at parents.
Laughs easily.
Does not seem to feel guilty after misbehaving.
Remains cool under pressure.
Takes a lot of risks in play.
Shy with unfamiliar adults.
Has trouble sleeping.

---

or prior behavior. A small proportion of the Q-sorts was administered by the examiner after the laboratory evaluation had been completed.

Most scientists work to satisfy two motives: to understand the small part of nature they selected to probe and to persuade colleagues that their novel inferences deserve both reflection and a little admiration. The strategies that serve these two motives need not be consistent. The conflict is especially clear in psychology because most social scientists have been trained to trust the standard statistical tools and to distrust inferences based on arbitrary divisions of distributions into terciles, quartiles, or deciles. This conviction rests on the assumption that every hypothetical construct associated with a distribution of scores should be regarded as a continuous property, an issue discussed in Chapter 2. We doubt the validity of this claim. We examined many scatterplots; and when we believed that an inference following an examination of extreme scores had theoretical significance, we described that result. However, we also wished to persuade our colleagues, and therefore we implemented and reported the results of standard statistical analyses.

Most of the time the magnitudes of our significant correlation coefficients were between .3 and .5. By comparison, in the general population the correlation between height and weight is about .4; between ingestion of sleeping pills and a reduction in insomnia about .3; between the Graduate Record Examination score and grades in graduate school, .22; between college grades and later job performance, .16; and between smoking and lung cancer, .08 (Meyer et al., 2001). These facts should be remembered as one reflects on our results.

Behavior

Videotapes of the laboratory session were coded for the number of spontaneous comments and smiles during the first 18 minutes of interaction, before the physiological measurements required the children to be quiet (the intercoder reliability for a randomly selected sample of 40 children was $r = .85$ for comments and .90 for smiles). In addition, an observer rated, on a 4-point scale, the degree of tension, uncertainty, and restrained affect, compared with a relaxed, spontaneous mood with the examiner, during the initial 18 minutes (an independent rating of the 40 cases by a second observer yielded a kappa of .76). Although this rating of inhibition was influenced primarily by the number of comments and smiles the child displayed, the judgment also took into account motor tension, restless movements of fingers or limbs, questions that indicated concern about the procedures, failure to look at the examiner, and speaking in a soft voice.

A rating of 4 (given to 21 percent of the children) described a child who was extremely quiet, rarely smiled, showed a great deal of motor tension, and spoke in a soft voice—the properties of an inhibited child. A rating of 3 (given to 14 percent) described a child who showed an average number of comments and smiles but displayed some other signs of uncertainty. A rating of 2 (30 percent) described a child who was generally quiet but showed no other signs of uncertainty. A rating of 1 (35 percent) described a child who was relaxed, spoke and smiled frequently, had a voice of average loudness, and showed no signs of motor tension—this is the profile of the uninhibited child.

Spontaneous smiles, especially those that involve the muscles of the eyes as well as the mouth (called *felt* or Duchenne smiles), are known to correlate with mood states. For example, adults who have not yet shown the symptoms of schizophrenia but will do so later in life display few spontaneous smiles in interviews with strangers (Dworkin et al., 1993). Spontaneous smiles become more frequent when depressed patients are in remission and when adults are recovering from a period of mourning (Schelde, 1998; Keltner and Bonanno, 1997).

Felt smiles appear to result from greater participation of the motor cortex in the right hemisphere, because most individuals show a larger aperture on the left side of the mouth, compared with the right, when this class of smiles occurs (Wylie and Goodale, 1988). These smiles seem to be released when a state of mild cognitive arousal is resolved—the classic example being the smile of the 3-month-old upon seeing a human face. These smiles do not occur prior to 2 months because younger infants usually do not relate an event in the perceptual field to their small store of acquired knowledge. One-year-olds smile when they master a motor task following effort—for example, taking their first steps. Previous assessments of our sample, at 4.5 and 7.5 years of age, and evaluation of other children had revealed that the frequency of smiles and spontaneous comments during interactions with an examiner differentiated temperamental groups (Kagan et al., 1999). The children who had been high-reactives were quiet and smiled infrequently, presumably because they were more uncertain in unfamiliar social settings.

As 11-year-olds interact with a stranger, smiles are most likely to occur when their mild uncertainty is reduced momentarily. Hence, children who rarely smile may not experience as complete a resolution of this state. We view the frequency of spontaneous smiles as an indirect index of the degree to which this resolution occurred during the first 18 minutes of the interview with the unfamiliar woman.

As Tables 4.3 and 4.4 show, the children who had been high-reactives were rated as more inhibited than those who had been low-reactive (F 2/232 = 9.8, $p < .001$). Twice as many high- as low-reactives were rated as extremely inhibited; twice as many low- as high-reactives were rated as uninhibited (chi-square (1) = 11.8, $p < .01$). One-third of high-reactives ($N = 22$), but only 14 percent of low-reactives ($N = 13$), received the highest rating for

Table 4.3 Means (and standard deviations) for rating of inhibition, comments, and smiles.

Variable	Low-reactives		Others		High-reactives	
	Boys	Girls	Boys	Girls	Boys	Girls
Rating of inhibition	1.9	1.8	2.3	2.2	2.7	2.6
	(1.0)	(1.0)	(1.0)	(1.1)	(1.1)	(1.2)
Comments	18.4	20.5	18.3	18.0	18.3	14.2
	(17.3)	(17.1)	(23.0)	(18.0)	(17.3)	(17.2)
Smiles	9.1	15.0	7.3	15.8	5.8	10.5
	(7.9)	(12.5)	(6.7)	(13.5)	(6.1)	(10.7)

Table 4.4. Percent of each temperamental group receiving varied ratings for inhibition with the examiner.

Rating	Low-reactives	Others	High-reactives
1	50	28	23
2	28	39	22
3	8	15	22
4	14	18	33

inhibition; one-half of low-reactives ($N = 46$), but only 23 percent of high-reactives ($N = 15$), received the lowest rating.

As we had found in the past, more high- than low-reactives were quiet, serious, spoke in a soft voice, sat stiffly in the chair, showed small movements of the hands or feet, asked questions about the safety of the procedures, and often looked away from the examiner. More low-reactives showed the opposite profile. The behaviors of the other children were intermediate between the high- and low-reactives: they smiled more than high-reactives but showed more signs of uncertainty in motor movements and spoke with a voice that was softer than low-reactives.

High-reactives smiled less often than low-reactives ($F\ 2/232 = 3.50$, $p < .05$). More high- than low-reactives had values in the lowest quartile of the distributions for both comments and smiles (25 versus 10 percent). More low- than high-reactives had values in the highest quartiles for both responses (15 versus 10 percent; chi-square $(1) = 4.2$, $p < .05$). The correlation between number of comments and a completely independent rating of "talkativeness at home" made by Christina Hardway was .53. Thus, the two infant temperaments predicted, to a modest degree, a posture of spontaneity or subdued restraint with an unfamiliar adult.

THE STABILITY OF BEHAVIOR

About one-half of this group of high- and low-reactive preadolescents had been seen at 4.5 and 7.5 years of age when a reliable rating of degree of inhibition (also on a 4-point scale) had been assigned to each child based on 60 minutes of interaction with a different unfamiliar female examiner. As with the 11-year assessment, more low- than high-reactives were rated as uninhibited, and more high- than low-reactives were inhibited. Seventy percent of low-reactives, compared with 13 percent of high-reactives, were rated as uninhibited in their behavior (receiving a rating of 1 or 2) at all 3 ages; 38 percent of high-reactives, but only 6 percent of low-reactives, received a rating at all 3 ages reflecting inhibited behavior (chi-square (1) = 21.3, $p < .001$). Four of every 10 high-reactives, but less than 1 of every 10 low-reactives, maintained an inhibited profile from 4.5 to 11 years of age. It was rare for a low-reactive infant to become a consistently inhibited child or for a high-reactive infant to become a consistently uninhibited child. The uninhibited profile was better preserved, we believe, because family and friends encourage sociability rather than shyness, and American children would rather be sociable than shy.

Further, more low- than high-reactives maintained an inclination to smile frequently from the second year to the 11-year assessment. Thirty-eight percent of low-reactives, 16 percent of high-reactives, and 29 percent of others had values for smiling above the median during the two assessments in the second year and at age 11, compared with 37 percent of low-reactives, 35 percent of high-reactives, and 37 percent of others who had values below the median at all 3 ages (chi-square (2) = 12.2, $p < .01$).

We also found a relation between a low resting heart rate and smiling. The children who smiled frequently at all 3 ages had lower heart rates than those who smiled infrequently. Figure 4.1 illustrates the mean heart period (inverse of heart rate) across the 6 phases of the challenging test for auditory acuity for high- and low-reactives who smiled either frequently or infrequently across the early and current assessments. The high-smiling children had lower heart rates.

Surprisingly, the 4-month temperamental classifications were a

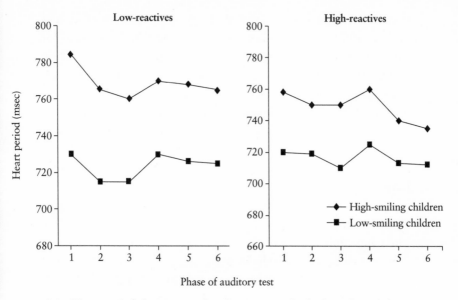

4.1 Heart period during test of auditory acuity for high-smiling and low-smiling children among low-reactives and high-reactives (at 14/ 21 months and 11 years). A higher heart period = a lower heart rate.

better predictor of contemporary behavior—and, as we shall see, of biology—than fearful behavior during the second year. Because too few low-reactives had high fear scores (only 7 percent), we could only evaluate this relation for the high-reactives and others. We found no consistent relation between level of fear in the second year and degree of inhibited behavior with the examiner at 11 years. These findings imply that fear scores in the second year reflect both infant temperament and life experiences. However, frequency of smiling during the second year did predict current smiling and laughter with the examiner. Thus, the 2-year-old's affective state with the examiner, whether joyful or serious, was a better predictor of behavior at age 11 than fearfulness to unfamiliar people, objects, and situations at the early age.

Children who smile frequently probably possess lower levels of cortical and sympathetic arousal (Kagan, 1971; Stifter, Fox, and Porges, 1989). Twin studies imply a substantial heritability for variation in smiling (Reppucci, 1968). The heritability of smiling and laughter in a large sample of monozygotic and dizygotic twin pairs approached .6, which is higher than the heritability of behavioral inhibition (Emde and Hewitt, 2001; see also Goldsmith and Campos, 1986). More high- than low-reactives approach each day with a serious frame of mind. Galen's melancholic and sanguine types were prescient (Niedenthal, Halbertstadt, and Innes-Kerr, 1999).

MATERNAL Q-SORTS

The temperament groups differed in the maternal Q-sort descriptions of shy versus sociable behavior (items 3, 4, and 24 for shyness; items 1, 8, 15, and 16 for sociability; see Table 4.2b). We computed the mean ranks across the 3 shy and the 4 sociable items ($r = -.52$ between these two values). As expected, low-reactives were described as more sociable ($F\ 2/219 = 7.71, p < .01$) and less shy ($F\ 2/219 = 10.4, p < .01$) than the other two groups, who had similar values. We subtracted the mean rank for sociability from the mean rank for shyness to construct a variable that reflected high sociability. The low-reactives were more sociable than the high-reactives and others ($F\ 2/219 = 11.76, p < .01$). About one-third of children in the high- and low-reactive groups were perceived by their mothers as possessing a personality in accord with theoretical expectations; only one-sixth were described in ways inconsistent with their infant temperament.

On the assumption that any Q-sort item given a rank of 1 reflected a salient trait, we asked what proportion of low- and high-reactives were given a rank of 1 for any of the 8 items that psychologists would regard as components of an extraverted profile (Q-sort items 1, 8, 9, 12, 15, 16, 26, and 28) and what proportion were given a rank of 1 for any of the 7 items that should be part of an introverted type (items 3, 4, 11, 20, 23, 24, and 25).

About 1 of every 2 low-reactives, but only 1 of 8 high-reactives (47 vs. 12 percent) received a rank of 1 for at least one of the extraverted items. But equal proportions of high- and low-reactive groups received a rank of 1 for one of the introverted items (chi-square (1) = 8.1, $p < .01$). It appears that low-reactive infants are especially biased to develop an extraverted profile.

Although the mothers' rankings of the child's shyness were correlated with subdued behavior with the examiner ($r = .28$) and with the rating of inhibition ($r = .39$), the magnitudes of these coefficients were modest and were influenced by a few children with extreme scores. The maternal Q-sorts were less discriminating of high- and low-reactives than the adolescent's behavior with the examiner, because some mothers were reluctant to ascribe a shy profile to their child. Forty-eight percent of high-reactives were very subdued with the examiner (z for comments and smiles $< .10$), but 38 percent of these children were described as sociable by their mothers ($z > .2$). By contrast, 37 percent of low-reactives were spontaneous with the examiner, yet only 10 percent in this group were described as shy.

Because the child's behavior with the examiner and the mother's descriptions were only modestly related, we considered both variables together. The odds ratio for high- over low-reactives was 5.6 to 1 when we compared children who were subdued with the examiner and described as shy with the children who were spontaneous and described as minimally shy. This odds ratio is larger than the ratio for each of the variables considered alone. Thus, high- and low-reactives were better differentiated when we combined their laboratory behavior with the mothers' descriptions.

Almost 80 percent of the current sample of mothers ($N = 189$) had filled out a questionnaire when their child was 7.5 years old. The 3-point scale read (1) my child rarely displays this behavior, (2) my child sometimes displays this behavior, (3) my child frequently displays this behavior. Four items dealt with shyness and 5 with disobedience. Similar items were present on the maternal Q-sort for 11-year-olds; hence, we asked about the preservation of the maternal descriptions of these two traits from 7 to 11 years.

The maternal ratings for shyness and disobedience were moderately stable ($r = .49$ for shyness; $r = .42$ for disobedience), but there was no relation, at either age, between the two traits, and more low- than high-reactives were described as both sociable and disobedient at both ages (chi-square (1) = 3.8, $p < .05$). We combined the maternal ratings for shyness and disobedience (at both ages) to create all possible groupings (based on a division at the median rating at each age). A majority of children (70 percent) preserved both traits; only 10 percent changed from high to low or from low to high for both variables across the 4-year interval from 7 to 11 years.

CHILDREN'S Q-SORTS

Most items on the children's Q-sort did not differentiate the 3 temperamental groups. However, mothers and children did have modest agreement with respect to shy or sociable behavior. The mean maternal rank for shyness items was correlated with the child's rank on a single item on the child's Q-sort "I am a shy person" ($r = .56, p < .01$). There was greater agreement between parent and child for sociability (21 percent of the child-mother pairs agreed that the child was sociable and minimally shy; only 7 percent of the pairs agreed that the child was shy and not sociable). Both mothers and children were reluctant to acknowledge shyness because it is a less desirable trait.

The low-reactives who described themselves as minimally shy and were described by their mother as highly sociable differed from the complementary group of low-reactives in their behavior with the examiner and in their heart rate. The sociable low-reactives were more spontaneous with the examiner and had lower baseline heart rates than the shy low-reactives ($t = 1.98, p < .05$ for behavior; $t = 2.41, p < .05$ for heart rate). But these differences were absent when we performed the same comparison with high-reactive children.

The statement in the child Q-sort that best distinguished high- from low-reactives was "Most of the time, I'm happy." Fifty-five percent of low-reactives, but only 32 percent of high-reactives and 44 percent of others, placed this item in ranks 1–3 (a value close to the median; chi-square (2) = 7.3, p < .05). Two-thirds of high-reactives did not regard a happy mood as one of their salient traits.

When we restricted the analysis to items given a rank of 1, more low- than high-reactives endorsed at least 1 of the 5 items reflecting an extraverted personality (43 vs. 27 percent). The most differentiating item was "I like to do risky things"; 12 low-reactives but only 5 high-reactives ranked this item as extremely characteristic of self. When we combined the mother and child Q-sorts with behavior with the examiner, more high- than low-reactives were inhibited with the examiner, were described as shy, and reported that they did not like new activities or playing with many children (chi-square (1) = 6.1, p < .01)

TRIADS PROCEDURE

We suspected that the children's Q-sorts might be subject to some distortion because most children would want to present themselves in a positive light and not admit to less desirable qualities, such as shyness or anxiety. For this reason, we used a procedure that we hoped would provide a more accurate appraisal of these properties. We asked the child to nominate 3 pairs of same-sex friends. One pair consisted of a friend who was very shy and one who was minimally shy; the second pair consisted of a friend who worried a lot and one who worried very little; the final pair referred to a friend who was very reluctant to take risks and one who took risks frequently. After the child made these 6 different nominations, she saw a row of 3 names on a monitor for a total of 35 trials. The 35 trios of names on the screen were randomly selected from the 6 peers nominated by the child. But on 15 of the 35 trials the child's own name appeared with two other friends and the child decided which name in each trio differed most from the other 2. The variable of interest was the number of times the child paired herself with a peer whom she had nominated earlier as shy or not shy, worried or not worried, a high risk-taker or a low risk-taker.

The data were generally similar for the children in the 3 temperamental groups. Surprisingly, low-reactive boys paired themselves with friends who did not take risks and who worried—a result that is the opposite of expectation. We computed a mean for the inhibited characteristics (shy, worried, and low risk-taking) and a mean for the uninhibited properties (not shy, not worried, and high risk-taking) and subtracted the latter from the former so that a high score reflected a tendency to pair oneself with more inhibited friends.

Low-reactive boys paired themselves with shy, low-risk, worrying peers; high-reactive boys paired themselves with uninhibited friends. Thus, the data from the triads procedure were less useful than we had hoped, in part because the children's selections were often based on qualities other than the one for which the friend had been nominated. The information on shyness that the children supplied was a less sensitive index of their past behavior and their interaction with the examiner than the mother's Q-sort descriptions.

BIRTH ORDER

The effect of birth order on personality has been a continuing source of debate, especially since Frank Sulloway (1996) suggested that first- and later-borns differ in their attitude toward legitimate authority. Therefore, we examined the possible influence of birth order on our data. Since half the children in our sample were first- and half were later-born, we were able to examine the behavioral differences between the two groups.

We found no major effect of birth order on any behavioral or biological measure, but we did find a significant interaction between birth order and gender for a few measures. The later-born boys from all 3 temperamental groups smiled more often than the first-borns (F 1/220 = 3.60, p = .05), and on their Q-sorts they reported liking new activities, roller coasters, and playing with different children (F 1/205 = 6.49, p < .01). This difference was not observed for girls, and there was no interaction of birth order with temperament for any biological variable. Thus, birth order was not an important correlate of our major measures.

To sum up: about 1 in 3 high-reactives (22 of 67 children) and 1 in 2 low-reactives (46 of 92 children) had developed social behaviors that were predictable from their infant temperaments. Only 8 high-reactives and 13 low-reactives developed a profile seriously inconsistent with expectations. This result is remarkable in light of the varied social experiences these children must have encountered over the past 11 years. The evidence suggests that high-reactive infants are biased to develop a shy, subdued style of social interaction; low-reactives possess a bias for a sociable, spontaneous personality.

PENNY AND PAUL

We close this section with brief profiles of two children who provide concrete representations of the two types. Penny (a pseudonym) was a low-reactive infant girl who was fearless in the second year and described by her mother as very sociable. Penny entered the laboratory talking, smiled 15 times during the first 5 minutes of meeting the examiner, and answered the examiner's questions with long sentences spoken in a voice high in energy. Paul, a high-reactive infant who was very fearful in the second year, entered the room cautiously, with a vigilant facial expression. After sitting down, he began to bite his fingers, pick at his face, and scan the room. When he finally spoke 8 minutes later, his voice was soft, his body tense, and he displayed a blinking tic. These two children, admittedly extremes, capture the quintessential behavioral differences between the high- and low-reactives during their initial encounter with the examiner. The most salient features of the high-reactives were a soft voice, motor tension, serious facial expression, and lack of emotional spontaneity in this unfamiliar setting.

Encounters with strangers are frequent events that cannot always be anticipated. The experience of interacting with an adult examiner in an unfamiliar laboratory room containing strange equipment should have created greater uncertainty in high- than in low-reactives. Most of the time, a single sample of laboratory behavior does not provide much insight into a person's deeper dispositions. But 18 minutes can be sufficient if the setting contains exactly the right incentives to engage an individual's stable temperamental qualities.

Biological Assessments

The next longer section of this chapter explains the rationale for our biological measurements and describes the profiles that separated the temperamental groups. These variables were chosen for two reasons: other scientists had discovered that they were associated with behaviors related to inhibited and uninhibited categories, and each was potentially influenced by amygdalar activity. The biological variables we studied fell into 6 categories:

- EEG power and asymmetry of activation
- Brain stem auditory-evoked potential
- Balance between sympathetic and vagal reactivity in the cardiovascular system
- Event-related potentials
- Startle and corrugator activity
- Anthropometry

Some readers may wish to read the summaries at the end of each of the sections and move on to the broader synthesis of behavior and biology presented in Chapter 5.

Rationale

The electroencephalogram (EEG) represents the synchronized activity of large numbers of cortical pyramidal neurons which, at any moment, have a dominant frequency of oscillation at particular sites. Variation in this frequency is correlated with particular psychological states. A state of mental and physical relaxation is usually, but not always, associated with more power in the alpha frequency band in frontal areas (8–13 Hz). Adult extroverts have more alpha power than introverts (Gale et al., 2001), and this frequency band increases in power from 6 to 11 years of age (Ray and Cole, 1985).

A state of psychological arousal, often due to cognitive work or emotion, is more often associated with more power in the higher frequency beta band (14–30 Hz), which could result from more intense volleys from the amygdala to the cortex via the basal nucleus of Meynert (Damasio, Adolphs, and Damasio, 2003). The relative amount of alpha and beta power at particular sites can be an indirect index of cortical arousal. In our context, the level of arousal might vary with the amount of concern over the procedures to be administered in this unfamiliar setting.

In addition, there are usually slight hemispheric differences in the amount of alpha power at frontal and parietal sites. Because alpha frequencies are associated with a relaxed psychological state, while higher frequencies are associated with an aroused state, the less alpha power at a particular site, the more likely that site is cortically active. The technical term for loss of alpha power is *desynchronization,* and an increase in desynchronization of alpha frequencies is a sign that a person has moved from a relaxed to a more aroused state.

When individuals are simply sitting quietly with no task to perform, about 50 to 60 percent show greater activation in the left frontal area (that is, less alpha on the left than on the right); between 20 and 25 percent show greater activation on the right. The remaining 15 to 30 percent show similar alpha power values in the right and left frontal areas. The asymmetry in activation in frontal areas is relatively independent of asymmetry of activation in parietal areas, even though each variable is moderately stable over a period of several weeks. Asymmetry of activation in the frontopolar area, anterior to the site where frontal asymmetry is usually assessed, appears to represent a more transient state and is assumed to reflect activity in the orbitofrontal cortex (Papousek and Schulter, 1998). There is reason to believe that the right frontopolar region contributes to the ability to appreciate humor in a joke or cartoon (Shamni and Stuss, 1999).

The left hemisphere is usually more active than the right when a person is in a happy or relaxed state, and individuals with greater activation in the left frontal area more often report sanguine moods, are biased to detect pleasant features in pairs of words, and show less anxiety than the smaller proportion who show greater activation on the right side (Davidson, Jackson, and Kalin, 2000; Davidson, 2003a; Fox, Calkins, and Bell, 1994; Schmidt, 1999; Sutton and Davidson, 2000). By contrast, an unpleasant experience, an unpleasant memory, and unfamiliar events are associated with greater activation in the right hemisphere (Piefke et al., 2003; Nagae and Moscovitch, 2002; Wacker, Heldmann, and Stemmler, 2003; Atchley et al., 2003; Ueda et al., 2003). Pictures of disfigured faces, unpleasant video clips, or aversive odors produced greater neural activity in the right hemisphere than in the left (Roschmann and Wittling, 1992; Pizzagalli et al., 1998; Tomarken, Davidson, and Henriques, 1990; Kline et al., 2000).

Musical excerpts that usually induce a sad feeling are more often accompanied by right frontal activation, while musical excerpts that generate a joyful feeling are more likely to generate left frontal activation (Schmidt, Trainor, and Santesso, 2003). Adults with severe melancholia showed greater power in the 13–30 Hz band in the right frontal area, compared with the left (Pizzagalli et al., 2002). Social phobics showed increased activity in the right dorsolateral prefrontal cortex when they thought they had to give a short speech while strangers watched (Tillfors et al., 2002).

Even infants show a relation between asymmetry and psychological state, for they display greater left than right frontal activation in response to film clips depicting a person in a happy mood (Davidson and Fox, 1982). Infants who cry when their mother departs temporarily show greater right frontal activation than those who do not cry (Davidson and Fox, 1989). Six-month-old infants who show extreme right frontal activation under baseline conditions have higher cortisol levels in the laboratory than other infants and, in addition, show a sad facial expression when a stranger approaches (Buss et al., 2003).

Further, uncontrolled shock produces a greater change in dopamine metabolism in the right, compared with the left, hemisphere of rats (Carlson et al., 1993; Roschmann and Wittling, 1992; Sullivan and Gratton, 2002; Ray and Cole, 1985). The right amygdala seems to play a special role in creating a state of uncertainty, for the amount of increase in delta power is greater in the right than in the left amygdala when squirrel monkeys are in a situation that should generate uncertainty (Lloyd and Kling, 1991). Normal adults given dextroamphetamine and shown faces with either an angry or a fearful expression showed increased activity in the right amygdala, as expected (Hariri et al., 2002).

A rat's retention of an acquired avoidance response relies more on the right than on the left amygdala (Coleman-Mesches and McGaugh, 1995). Rats exposed to a male cat for 5 minutes—a stressful event for a rat—showed greater potentiation of neural transmission from the central nucleus to the central gray and from the ventral angular bundle to the basolateral nucleus of the amygdala on the right, compared with the left, side one hour following the stress (Adamec, Blundell, and Collins, 2001).

Adults secrete more cortisol and show greater increases in blood pressure when they view an emotionally arousing, aversive film in the left visual field (that is, with the right hemisphere) than when they view the same film in the right visual field, despite no difference in subjective emotion between the right or left visual field presentations (Wittling and Pfluger, 1990). Adults who watched both emotional and neutral films during separate sessions, and who were later given a surprise test of recall memory for the films, showed a high positive correlation ($r = .93$) between the amount of information recalled from the emotional films and PET activity in the right amygdala. Greater activity in the right amygdala should lead to greater EEG activation—loss of alpha—in the right hemisphere due in part to ipsilateral cholinergic projections from the amygdala to the cortex via the basal nucleus of Meynert (Cahill, 2000; Cameron and Minoshima, 2002).

The asymmetry in cortical activation could be influenced by inherent differences in excitability within each hemisphere, or by differential input from limbic or brain stem sites to the cortex, or both. A number of subcortical structures, including the amygdala, thalamus, and locus ceruleus, project to the cortex and modulate its activity (Baumgarten, 1993; Lopes da Silva et al., 1973; Wieser and Siegel, 1993). Neural activity in any of these sites, and especially the amygdala, would desynchronize alpha activity in the cortex, perhaps asymmetrically (Kapp, Supple, and Whalen, 1992).

If the right and left amygdala vary in activity, there should be an asymmetry of activation in the cortex. Visceral activity in the body—cardiovascular system, gut, and muscle—is transferred to the amygdala, and there is reason to believe that this bodily input is greater to the right amygdala (Wittling, 1995). Therefore, children with a greater visceral response to unfamiliar events should have a more active right amygdala and greater desynchronization of alpha frequencies in the right frontal area. These children would be classified as right-frontal-active. A child who had less visceral arousal might be biased to show left frontal activation.

No one is consciously aware of their asymmetry of activation. In one study, one angry face was paired with a 1-second burst of aversive white noise; a different angry face as well as two neutral faces were not accompanied by the unpleasant white noise. These adults were then scanned under two conditions. In the first condition, a face appeared for only 30 msec, followed by a masking stimulus that prevented the subject from being consciously aware of the expression of the face on the screen. In the second condition, the faces were presented long enough for the subject to perceive the facial expression.

The right amygdala was more active than the left when the adults could not perceive the angry face that had been paired with the aversive white noise. But the left amygdala was more active than the right when the subject could perceive the face. This evidence, along with others, implies the possibility of two distinct brain states linked to an aversive experience. One is unavailable to consciousness and more closely related to right amygdalar activity and therefore to right hemisphere activation in the EEG (Dolan, 2000a,b).

One important and as yet unresolved issue is whether right or left frontal asymmetry reflects a stable trait or a transient state. The evidence supports both views. An asymmetry of activation that is related to an individual's temperament implies a stable property (Fox et al., 1995; Field et al., 2002; Schmidt et al., 2003). Four-year-olds who were shy with 3 other unfamiliar children in a laboratory playroom showed right frontal activation, while those who were sociable showed left frontal activation (Henderson et al., 2002). A similar result was found in 9-year-olds (Rickman, 1998).

Adults who reported being shy and minimally sociable showed right frontal activation, while extroverts, with the opposite persona, showed left frontal activation (Schmidt, 1999). Women who were serious about restricting their eating also showed right frontal activation (Silva et al., 2002). One study found the direction of asymmetry in the frontal area when subjects were awake to be almost perfectly correlated with the direction of asymmetry when the subjects were asleep (Schmidt, Trainor, and Santesso, 2003).

This fact suggests that a bias for left or right hemisphere activation may be an inherent property of an individual brain.

But asymmetry of activation can also reflect a transient state (Davidson, 2003b). Fox and colleagues reported greater right frontal activation in infants watching the approach of a stranger (Fox and Bell, 1990; Fox, 1989), and the direction of frontal asymmetry in adults was correlated with transient facial expressions for varied emotional states (Ekman, Davidson, and Friesen, 1990). Adolescents showed left frontal activation while listening to music they liked but right frontal activation to music they did not like (Altenmuller et al., 2002). Adult EEG data gathered on 4 separate occasions suggested that 60 percent of the variance in asymmetry reflected a stable trait, while 40 percent was attributable to the specific time of the assessment (Hagemann et al., 2002).

The entire corpus of evidence suggests that differential activation of the right and left hemisphere, especially in frontal areas, is associated with variation in affect states and the initial reaction to unfamiliar events. For these reasons, we expected that more high-reactives would show right hemisphere activation, and more low-reactives would show left hemisphere activation. This expectation is supported by the fact that 4-month-old high-reactive infants observed in the laboratory at the University of Maryland had greater right frontal activation at 14 months and were more likely to be behaviorally inhibited than other infants (Calkins et al., 1996).

We also had reason to expect that the direction of the asymmetry of activation would be related to absolute power in the alpha and beta bands. Heightened emotional arousal in the laboratory should be accompanied by activity in the higher frequencies, and perhaps by a more distinct asymmetry of activation (Schaul, 1998; Moelle et al., 2002). Thus, children with more beta relative to alpha power under resting conditions, who presumably are under higher levels of cortical arousal, might be biased to display right hemisphere activation. It is relevant that adult introverts have less alpha and more beta power in frontal areas than extroverts (Trayn,

Craig, and McIsaac, 2001; Gale et al., 2001). One group of shy 4-year-olds with right frontal activation had high levels of alpha power in the left frontal area and therefore were right-frontal-active (Fox et al., 1995). Thus, we were prepared to examine the relations between the direction of asymmetry of activation and the balance between alpha and beta power.

We measured alpha and beta power and asymmetry of alpha power at frontal and parietal sites under 3 conditions. During the first and third minutes, the child was asked to sit quietly with eyes open. During the second and fourth minutes, the child was asked to close his eyes. During the fifth minute, the child was asked to silently prepare a speech describing what he was thinking about on the way to the laboratory and then to stand up and recite that narrative. The expectation was that the children who had been high-reactive infants would show greater increases in beta power to the challenge of preparing the speech, and greater right hemisphere activation.

EEG electrodes were placed according to the international 10/20 system at F_3, F_4, P_3, P_4, A_1, A_2, referenced to Cz for data collection. Vertical eye movements were recorded from Ag/AgCl electrodes placed super- and suborbitally on the left eye, and horizontal eye movements were recorded from the electrodes placed at the outer canthus of each eye. Data were sampled at 1,000 Hz, and signals were band-pass filtered between .08 and 100 Hz. EEG and eye movement data were resampled at 250 Hz and re-referenced to average linked earlobes offline (A1 plus A2/2). The wave forms were inspected visually for eye movement or muscle artifacts. When an artifact was identified on any channel, the time was marked, and all data on all channels were excluded for that time interval.

The spectral power density (uv/RMS) was computed using a fast Fourier transformation on 2.048 seconds of artifact-free data. Epochs overlapped by 1.024 seconds and were passed through a Hamming window to reduce error in spectral power estimates at the beginning and end of each episode. Alpha power was defined as the sum of spectral power between 8 and 13 Hz; beta power as the sum of spectral power between 14 and 30 Hz. The spectral power estimates were log-transformed to normalize the distribution. Asymmetry of activation was computed by subtracting log alpha power on the left from log alpha power on the right; hence, a positive value reflected less alpha power and therefore greater activation on the left side.

Findings

Asymmetry. There were no significant differences among the 3 temperamental groups in alpha or beta power or frontal asymmetry (Table 4.5). However, high-reactives had greater right hemisphere activation at parietal sites than low-reactives (eyes-open condition, $F = 3.91$, $p < .05$) and were more likely to have asymmetry values in the lowest quartile of the distribution of difference scores (reflecting right hemisphere activation). Further, the high-reactives who, in addition, had been highly fearful in the second year were more likely than the low-reactives or others with equally high fear scores to show right hemisphere activation at frontal sites under at least 1 of the 3 conditions (63 versus 41 percent, chi-square (1) = 5.9, $p < .05$) (see Figure 4.2). High-reactives who were highly fearful in the second year (fear score > 3.0) and preserved an inhibited persona through age 11 were more likely to be right-frontal-active if a boy and right-parietal-active if a girl, compared with the high-reactive/high-fear children who were not inhibited at 11 years (chi-square (1) = 7.0, $p < .01$). Thus, high-reactive children who were fearful in the second year and remained inhibited through age 11 were more likely to show right frontal activation than high-reactives who developed inhibition later or who lost their inhibited persona during childhood.

Power. The low-reactives had a distinctive profile of high alpha and low beta power at frontal sites (across all 3 conditions; 33 per-

Table 4.5. Values for major EEG variables for low-reactives, high-reactives, and others.

Variable	Low-reactives	Others	High-reactives	F or chi-square
Alpha power				
Speech	3.14	3.06	3.10	
Eyes closed	3.58	3.53	3.62	
Eyes open	3.12	3.05	3.10	
		(SD = 0.4)		
Beta power				
Speech	3.15	3.34	3.32	
Eyes closed	2.94	3.02	3.05	
Eyes open	3.17	3.36	3.28	
		(SD = 0.5)		
Frontal asymmetry (right-left)*				
Speech	.05	.05	.04	
Eyes closed	.04	.08	.05	
Eyes open	.02	.04	.03	
		(SD = .18)		
Parietal asymmetry (right-left)				
Speech	−.02	.00	−.08	
Eyes closed	−.08	.00	−.10	
Eyes open	−.03	−.03	−.13	2/221 = 3.91, $p < .05$
		(SD = .18)		
Left frontal/left parietal (eyes open)	29%		13%	
Left frontal/right parietal (eyes open)	17%		35%	Chi-square (1) = 14.0, $p < .01$

*Positive values indicate greater activation on the left side.

cent were high alpha/low beta, compared with 10 percent of high-reactives and others). By contrast, 23 percent of high-reactives and others showed a profile of low alpha/high beta power, compared with 10 percent of low-reactives (chi-square (1) = 16.0, $p < .01$). One of every 4 low-reactive boys combined 3 factors: (1) low beta

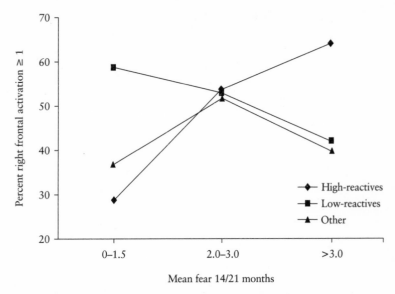

4.2 Percent of each temperamental group with right frontal activation under at least one condition as a function of temperament and fear score in the second year.

power at both frontal and parietal sites, (2) left frontal activation, and (3) a rating of extremely uninhibited behavior with the examiner (rating of 1), compared with only 1 of 20 high-reactive boys.

An important reason for accommodating to a child's temperament is that the functional relations between measurements can vary for children belonging to distinct groups. For example, when low-reactives were quiet with the examiner (standard score < .00), they showed greater alpha than beta power under the eyes-open condition. However, when high-reactives were quiet, they showed greater beta than alpha power (chi-square (1) = 4.1, p < .05). These patterns suggest that absence of spontaneous conversation among low-reactives is associated with low cortical arousal; the same behavior among high-reactives is associated with high cortical arousal. Put differently, when low-reactives are relaxed, they are quiet; when high-reactives are aroused, they become quiet.

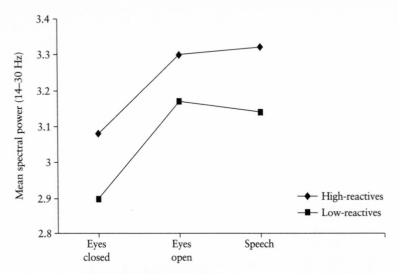

4.3 Mean spectral power (14–30 Hz beta) under three conditions for high- and low-reactive children.

This difference supports the claim in Chapter 1 that different brain states can produce a similar behavior.

Reaction to challenge. The challenge represented by the examiner's request to prepare a speech provoked a larger gain in frontal beta power among high- than low-reactives (see Figure 4.3). Thirty-one percent of high-reactives but only 16 percent of low-reactives showed large increases in beta power (> .15 microvolts). Fifty percent of low-reactives but 26 percent of high-reactives showed smaller increases in beta power from the eyes-open to the speech condition (chi-square (1) = 7.0, p <.01). And the high-reactives who displayed the largest increase in beta power smiled least often during the second year ($r = -.3$, p <.05), as well as contemporaneously.

The correlation between beta power at frontal and parietal sites—an index of coherence of beta power—was equally large for high- and low-reactives under the eyes-open condition ($r = .5$) but lower for high-reactives under the speech condition ($r = .3$ for high-reactives; $r = .5$ for low-reactives). The decreased coherence of beta power among high-reactives, which was clearer on the left side, was due to a large gain in beta power (and a corresponding loss of alpha power) at frontal sites to the speech challenge. This result is concordant with the fact that social phobic patients anticipating a speech challenge showed a greater increase in heart rate and increased right frontal activation than controls (Davidson et al., 2000).

Asymmetry of activation and behavior. The children who spoke frequently with the examiner, across all temperamental groups, showed moderate rather than extreme left frontal activation (quartile 3 of the distribution of asymmetry values). Low-reactive boys who were spontaneous and described by their mother as sociable had smaller frontal asymmetry values. The high-reactive boys who were subdued and described as shy had extreme asymmetry values in either the right or the left hemisphere ($z \pm .5$).

This surprising result motivated a comparison of the children who had large vs. modest asymmetry values. We selected an asymmetry value at frontal or parietal sites of $\pm .3$ or larger to define a large asymmetry group ($N = 73$); a z score of $-.29$ to $+.29$ at both sites defined the more frequent, small asymmetry group ($N = 115$). High-reactives with large asymmetries and high beta power were subdued with the examiner; low-reactives with the same asymmetry and beta power profile were spontaneous. If we assume that high beta power and a large asymmetry reflect a heightened state of cortical arousal, it appears that the behavioral correlates of that brain state vary with the child's temperament. Highly aroused low-reactives talk and smile frequently; aroused high-reactives are subdued. This is a second instance in which the relation between two measurements varied with temperament.

Further, the small number of high-reactives who had been highly fearful in the second year and behaviorally subdued at age

11 displayed an extreme degree of left frontal activation. Nine of every 10 low-reactives who combined low fear in the second year with spontaneity with the examiner showed moderate left activation. It is relevant that Buss et al. (2003) also found a nonlinear relation between the degree of right frontal activation in 6-month-old infants and cortisol level in the laboratory. Only the infants with an extreme degree of right frontal activation had high cortisol levels; infants with moderate right or left frontal activation had similar, and lower, levels. These facts imply that the magnitude of frontal asymmetry should not be treated as a continuous measure. Left frontal activation does not always reflect a relaxed mood or the tendency to approach new events (Davidson, 2003a).

Q-Sorts and the EEG. Several unexpected relations between the Q-sorts and EEG profiles warrant mention. We noted that low-reactives were more likely than high-reactives to describe themselves as "happy most of the time." The 20 low-reactive boys who claimed they were happy displayed a distinct cluster of biological properties different from the 21 low-reactive less happy boys. The former were left-active at both frontal and parietal sites and had significantly less alpha power and a lower baseline heart rate ($t =$ 2.25, 2.00, 2.10, $p <.05$).

A child could rely on his current feeling tone or on life conditions as he decided where to place the "happy" item. It is possible that the low-reactive boys who felt happy were relying on their feeling tone because their lower beta power and heart rate imply lower cortical and autonomic arousal. The other children may have relied more on a cognitive analysis of their life conditions as they made the same decision.

The association between left parietal activation and a sanguine mood is supported by the children's ranking of two other items: "I worry too much about making mistakes" and "I wonder a lot about what other children think of me." As expected, the low-reactives were a little less likely than the high-reactives to endorse those two items. However, the children who denied worrying about mistakes or peer opinion (across the whole sample) were more often left-active at parietal sites (chi-square (1) = 4.6, $p < .05$). The high-reactive girls who worried about both mistakes and peer opinion showed right parietal activation. However, low-reactive girls with the same self-descriptions did not show this biological property. Once again, similar self-descriptions had different biological correlates in high- and low-reactives.

Perhaps the most intriguing relation between the child's self-reports and asymmetry involved the item "I feel really bad if one of my parents said I did something wrong." Every high-reactive boy who was right-active at frontal and parietal sites ranked this item as more descriptive than the total sample, compared with only one-third of the high-reactive boys who were left active at both sites ($p < .05$, by the exact test). Although this relation between admission of guilt and right hemisphere activation was restricted to high-reactive boys, it deserves attention because some children inherit a temperament that biases them to experience frequent but unpredictable dysphoric feelings, which they attribute to violations of an ethical standard. The fact that high-reactive boys with right hemisphere activation reported more frequent guilt over disobedience supports this hypothesis.

These high-reactive, right-hemisphere-active boys also reported that they did not like "going on roller coasters in amusement parks." Only 23 percent of right-active, high-reactive boys endorsed this item, compared with 67 percent of left-active, high-reactive boys who liked roller coasters (chi-square (1) = 6.4, $p < .01$). The sensory feelings generated by roller coasters is more fully represented in the right than in the left hemisphere. It is possible that children with a high-reactive temperament, who have a more active right hemisphere, might experience the sensations generated by roller coasters as aversive. The absence of this relation among low-reactive boys who were right-hemisphere-active suggests that asymmetry alone is not sufficient to produce the aversion; it has to be combined with a high-reactive temperament.

The mother's description of her child as having "a high level of energy" was also correlated with EEG profiles. Although low-reactive boys and high-reactive girls most often received this description, these two groups displayed different EEG profiles. The low-reactive energetic boys combined low beta power with left hemisphere activation at both frontal and parietal sites (58 percent); only 22 percent of high-reactive girls who were equally energetic showed that pattern. Thus, a maternal perception of "high energy" was assigned to children with different biological features, probably because the behavioral signs parents relied on had different meanings for the two groups.

Earlier behavior. About one-half of the current sample of high- and low-reactives (31 high-reactives and 36 low-reactives) had been seen in our laboratory at 4.5 and 7.5 years. The number of comments and smiles displayed toward the examiner during these two assessments were coded reliably from the videotapes of the 60-minute batteries, and we computed a mean standard score for comments and smiles across the 3 ages (4, 7, and 11 years). As expected, more low-reactives were spontaneous (mean $z > .00$); more high-reactives were subdued (mean $z < .00$) (chi-square (1) = 7.9, $p < .01$). Sixty-one percent of high-reactives were subdued at all 3 ages; 33 percent of low-reactives were spontaneous during all 3 evaluations.

The more interesting fact is that a consistently subdued style, across all 3 ages, was associated with right parietal activation in high-reactive girls, but with left parietal activation in high-reactive boys. Specifically, 89 percent of high-reactive girls who were consistently subdued were right active parietally, compared with 40 percent of low-reactive equally subdued girls. Sixty-one percent of high-reactive boys who were subdued were left parietal active, compared with 33 percent of low-reactive subdued boys. Similar behaviors were associated with different biological features in boys and girls possessing the same temperament.

Heller's hypothesis. Finally, these data permit a partial evaluation of Heller's (1990) imaginative suggestion that extroverts should show a combination of left frontal activation for alpha frequencies and greater beta power in the right parietal area; introverts should show right frontal activation combined with higher beta power in the left parietal area. We evaluated this idea by comparing two groups. Group 1 showed left frontal activation and greater beta power on the right than on the left at parietal sites ($N = 67$). Group 2 showed the opposite profile of right frontal activation and greater beta power in the left parietal area ($N = 20$). These two groups did not differ on any behavior or combination of behavioral variables. Neither infant temperament, behavior with the examiner, nor maternal or child Q-sorts differentiated these two groups. These results cannot refute Heller, but they do not support her creative predictions.

Summary

The data were in partial accord with our predictions but contained some surprises. High- and low-reactives differed in EEG asymmetry and power in ways that are consonant with theory and the results of other laboratories. High-reactives were more likely (1) to be right hemisphere active at parietal sites; (2) to combine high fear in the second year with right frontal activation; and (3) if still subdued at age 11, to show right hemispheric activation. The low-reactive boys combined moderate left activation with low beta power; high-reactive boys combined extreme left frontal with right parietal activation and high beta power. The robustness of this last result is supported by data gathered on a group of 8- and 11-year-old German children. The anxious girls showed greater right frontal activation than control girls, but the anxious boys had greater left frontal activation than control boys (Baving, Laucht, and Schmidt, 2002). The similarity between the results of this study and our own invites the suggestion that the relation between temperament and direction of asymmetry interacts with gender, at least in children.

Finally, EEG asymmetry was related to the child's affective characteristics (as revealed in the Q-sorts), but the exact relation depended on temperament and gender. This fact, to be supported by later evidence, implies that the biology that underlies EEG asymmetry values should not be regarded as forming a continuum. Consistently subdued high-reactive boys showed extreme left frontal or parietal activation, while spontaneous, low-reactive boys showed moderate left frontal activation. Thus, the relation between magnitude of asymmetry and spontaneity was nonlinear, a common feature of functional relations in the life sciences.

BRAINSTEM AUDITORY-EVOKED POTENTIAL

Rationale

The brain stem auditory-evoked response (BAER), elicited by a series of clicks delivered through earphones, is relevant to our investigation of temperament because variation in the magnitude and/or latency of the fifth wave form in the BAER—Wave 5—which originates in the inferior colliculus, differentiates among personality and clinical categories. The peak of the fifth wave in the BAER is believed to represent the termination of the lateral leminiscus on the inferior colliculus (Moller et al., 1994; see Figure 4.4). The magnitude of Wave 5 increases across the first 5 years of life and then decreases gradually until adulthood (Ornitz et al., 2001). Wave 5 magnitudes are reliable over time (Stelmack, Knott, and Beauchamp, 2003), and introverts, compared with extroverts, have a faster latency to the peak of Wave 5 (Stelmack, 1990; Bullock and Gilliland, 1993; Swickert and Gilliland, 1998).

A positive relation between a high-reactive temperament and the magnitude of Wave 5 is a reasonable idea because the amygdala influences the inferior colliculus through projections to the central

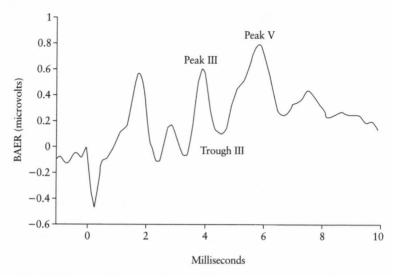

4.4 Illustration of the wave forms of the BAER. (Adapted from Chiappa, 1997.)

gray and the locus ceruleus; the projection is direct in the moustache bat (Marsh et al., 2002).

Many behavioral biologists regard the inferior colliculus as a component of a fear circuit because electrical stimulation of the rat inferior colliculus is followed by bodily freezing and reluctance to enter the brightly lit areas of the elevated maze (Brandao, Coimbra, and Osaki, 2001; Pandossio et al., 2000). The fact that more intense stimulation of the colliculus is necessary to produce the same duration of freezing if the amygdala has been lesioned (Macedo et al., 2002; Maisonnette et al., 1996) implies that the amygdala primes the inferior colliculus. Further, a conditioned stimulus signaling electric shock produces a large evoked potential in the inferior colliculus (Brandao, Coimbra, and Osaki, 2001; Brandao et al., 2003). This evidence suggests that children with a more excitable amygdala should show a larger Wave 5 (Brandao et al., 1994; Klepper and Herbert, 1991; Meller and Dennis, 1991; Price and Amaral, 1981). If high-reactives had a larger Wave 5 than low-reactives, our major hypothesis would be supported.

Each child was screened for ear health prior to measuring the BAER. The screening included a medical history from the parent, visual inspection of the ear, tympanometry, and a standard test of auditory acuity (a 2-interval, 2-alternative, 4-choice staircase procedure at 500 Hz and at 2,000 Hz adopted from Buus, Florentine, and Paulsen, 1997). Through earphones, the child heard a single tone. He was told that two lights in front of him would be lit in succession, but the tone would be heard when only one of the lights was lit; when the other light came on, there was no tone. After both lights had lit, the child had to indicate which light was associated with the presentation of the tone. The tones became softer following a fixed schedule. The few children with impaired hearing were excluded from this analysis.

The BAER was collected with Ag/AgCl electrodes at Cz, A_1, and A_2. The recording epoch was 37 msec (10 msec were utilized for analysis). Wave forms were amplified with an isolated bioelectric amplifier, and analog band-pass filtered at .08 to 3,000 Hz. The data were digitized at 10,000 Hz and digitally refiltered at 30 to 3,000 Hz. The peaks and troughs of Waves 1, 3, and 5 from the reference electrode ipsilateral to the stimulated ear, and Waves 3 and 5 from the reference electrode contralateral to the stimulated ear, were identified with a computer algorithm and checked manually by a coder unaware of the child's behavior. The child heard a series of clicks in the right ear through earphones. There were two blocks of 2,200 clicks of alternating polarity (.1 msec clicks; 70 dbspl for Block 1, 80 dbspl for Block 2) presented at a rate of 27 per second. Each block of clicks lasted approximately 1.15 minutes, and there was a 1-minute break between presentation of the first and second blocks. The variable of interest was the difference (in microvolts) between the trough of Wave 3

Table 4.6. Contralateral Wave 5 values (and standard deviations) at 70 and 80 db (microvolts).

	Low-reactives		Others		High-reactives	
dB	Boys	Girls	Boys	Girls	Boys	Girls
70	.62 (.22)	.65 (.22)	.68 (.22)	.71 (.18)	.76 (.23)	.74 (.26)
80	.70 (.26)	.69 (.12)	.74 (.20)	.82 (.20)	.80 (.22)	.73 (.28)

and the peak of Wave 5 for both ipsilateral and contralateral values at each loudness level.

Findings

The magnitude of Wave 5, was, as expected, larger on the contralateral than on the ipsilateral side and larger to 80 than to 70 dB (see Table 4.6). A repeated measures analysis of variance (2 loudness levels and ipsilateral and contralateral scores, and the 3 independent temperament groups) yielded a significant effect for the within conditions (F 3/549 = 69.7, $p < .001$), and for temperament (F 2/183 = 2.42, $p = .09$). The important result was that high-reactives had larger Wave 5 values than both low-reactives and others (F 2/198 = 4.41, $p < .05$). Twice as many high- as low-reactives had Wave 5 values in the top tercile of the distribution. This temperamental difference was most significant when Wave 5 was measured contralaterally at 70 dB; for that reason we concentrate on this variable for the remainder of the discussion (see Figure 4.5).

The possibility that Wave 5 magnitudes might reflect a stable temperamental feature is affirmed by the fact that high-reactives with the highest motor activity scores at 4 months had the largest Wave 5 values 11 years later ($r = .38$, $p < .01$, within high-reactives). Very high motor activity scores in infants were often accompanied by an obvious arching of the back, which is mediated by the central gray—a target of amygdalar projections. Some high-reactives arched as many as 20 times during the battery. The

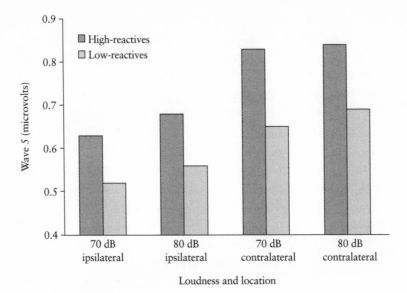

4.5 Mean amplitude of Wave 5 for high- and low-reactive children at two loudness levels and ipsilateral and contralateral sites.

few low-reactives who had moderately high motor activity scores rarely displayed this response.

Wave 5 was also related to extreme inhibition displayed seven years earlier when each child played with two unfamiliar peers in the laboratory setting. About 60 percent of the current sample had been observed at 4.5 years of age in a 45-minute play session with two other unfamiliar children of the same sex and age while all 3 mothers sat on a couch in a playroom. Each child was classified reliably as inhibited, uninhibited, or neither, based on her play behavior with the other two children, as well as her behavioral reactions to two unfamiliar events occurring after the play session.

As we noted in Chapter 1, more high- than low-reactives were inhibited; more low-reactives were uninhibited during this session. Of greater importance is the fact that the high-reactives who had been inhibited during this session had larger Wave 5 values at age 11 than high-reactives classified as uninhibited at 4.5 years of age.

High Wave 5 values were also related to absence of smiling. The small group of children (15 percent) who rarely smiled during the laboratory visits in the second year, as well as at age 11 (17 high-reactives and 8 low-reactives) had higher Wave 5 values than the 15 percent of the sample who smiled frequently at all 3 ages (8 high-reactives and 22 low-reactives) ($t (53) = 2.00, p < .05$).

Wave 5 was also linked to a sanguine mood. The children with small Wave 5 values and left activation at frontal and parietal sites—across all children—were most likely to describe themselves as happy. When we combined the "happy" item with a related item from the mother's Q-sort ("My child laughs easily"), 36 percent of all children who placed happy in ranks 1–4 and whose mothers placed laughing easily in ranks 1–7 had Wave 5 values in the lowest tercile. By contrast, only 12 percent of children with ranks for "happy" greater than 4 and for "laughing easily" greater than 7 had especially small Wave 5 values (chi-square (1) = 5.0, $p < .05$)—a ratio of 3 to 1 (see Table 4.7).

Because both Wave 5 and right parietal activation differentiated high- and low-reactives, we examined both variables together. Twenty-four percent of low-reactives but only 8 percent of high-reactives combined low Wave 5 values (lowest tercile) with left parietal activation. By contrast, 33 percent of high-reactives but only 20 percent of low-reactives combined high Wave 5 values (highest tercile) with right parietal activation (chi-square (1) = 6.9, $p < .05$). The ratio of high- to low-reactives who combined right parietal asymmetry with a large Wave 5 was 1.8 to 1 (expected ratio is .74). The ratio of low- to high-reactives who combined left parietal asymmetry with a small Wave 5 was 4.8 to 1 (expected ratio is 1.3, chi-square (1) = 10.5, $p < .01$).

About 1 of 4 high-reactives, but only 1 of 14 low-reactives, combined high Wave 5 values (highest tercile), high beta power values (greater than the median), and right parietal asymmetry. But

Table 4.7. Percent of children in each tercile of Wave 5 who are sanguine versus not sanguine.

Wave 5 tercile	"Happy" ranks 1–4 "Laughs easily" ranks 1–7	"Happy" rank > 4 "Laughs easily" rank > 7
1 (Low)	36	12
2 (Moderate)	36	42
3 (High)	28	45

Chi-square (2) = 5.0, $p < .05$

surprisingly, the high-reactives with this combination of features did not differ from the remaining high-reactives in their behavior with the examiner, early fear score, or maternal descriptions.

Summary

Variation in the Wave 5 magnitude separated high- and low-reactives better than any of our biological measurements—EEG asymmetry, sympathetic reactivity, or magnitude of the event-related potential. Because Wave 5 magnitudes were theoretically more closely linked to amygdalar excitability than the other variables, this result provides indirect support for our interpretation of the bases for high and low reactivity. However, Wave 5 values were not associated with early fear scores, contemporary social behavior with the examiner, or maternal reports of shyness. These facts invite the suggestion that the brain state indexed by Wave 5 magnitudes, when considered alone, has little relation to well-practiced social behaviors. Small Wave 5 values might index a brain state that contributes to a sanguine mood, but not to a shy or sociable posture with others.

SYMPATHETIC REACTIVITY IN THE
CARDIOVASCULAR SYSTEM

Rationale

Study of the relation between behavior and activity in the sympathetic nervous system has a long tradition in psychology. The ori-

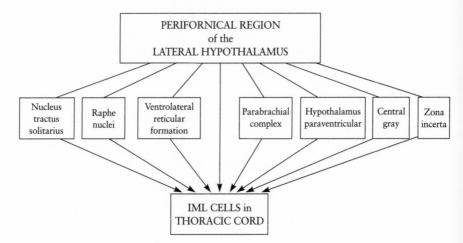

4.6 Projections from the perifornical region of the lateral hypothalamus. (Adapted from Loewy, 1990a.)

gin of the sympathetic nervous system can be traced to the hypothalamus, which projects first to the rostral ventrolateral nucleus of the medulla, then to the interomediolateral column of the spinal cord, and from there to a variety of sympathetic targets. The projections from the intromedialateral column on the right side control heart rate, while those on the left control blood pressure. The parasympathetic system originates in the dorsal motor nucleus of the medulla, which projects, in turn, to the nucleus ambiguus, and from there to various targets (Figure 4.6).

Cardiovascular activity is influenced by many brain centers, including the central nucleus of the amygdala, parabrachial nucleus, and lateral hypothalamus. We gathered systolic and diastolic blood pressure and heart rate under relaxed sitting and standing postures, as well as heart rate change to two mild cognitive stressors. One stress was the examiner's request to the child to prepare a speech; the second challenge was the test for auditory acuity.

We also quantified the ratio of high- to low-frequency power in the cardiac spectrum while the child lay supine on a mat. The spectral analysis uses the frequency distribution of the beat-to-beat differences in a sample of resting heart rate under regular respiration. A Fourier analysis of the distribution usually reveals two peaks in the distribution. The higher-frequency peak at about .2 Hz represents the influence of breathing on heart rate, which is vagal in origin. The lower-frequency peak, .05–.15 Hz, reflects both sympathetic and parasympathetic influences on heart rate, due primarily to slower cyclical oscillations in blood pressure and body temperature (Figure 4.7). It is believed that the magnitude of the low-frequency peak is due, in part, to the briskness of the baroreceptor reflex in the carotid sinus which responds to changes in blood pressure (Loewy, 1990a,b).

The central assumption in our analyses was that greater relative power in the lower-, compared with the higher-, frequency band reflected less vagal, and greater sympathetic, tone in the cardiovascular system. The specific variable of interest was the ratio of the amount of power in the high- compared with the low-frequency band; a low score reflected greater sympathetic influence. It is relevant that patients with panic disorder or PTSD have higher heart rates and more power in the lower-frequency band than controls (Cohen et al., 2001).

A small number of investigators measure variation in skin temperature, which is a function of degree of constriction or dilation of capillaries and sympathetically innervated arteriovenous anastomoses (the place where arterial and venous blood mix) in select body sites (see Figure 4.8). The sympathetic fibers to the hand are carried by the median and ulnar nerves; the median nerve serves the index, middle, and proximal half of the ring finger; while the ulnar nerve serves the other side of the ring and the small finger. The largest group of sympathetic neurons innervating the fingers, which is noradrenergic, causes vasoconstriction and cooling of the

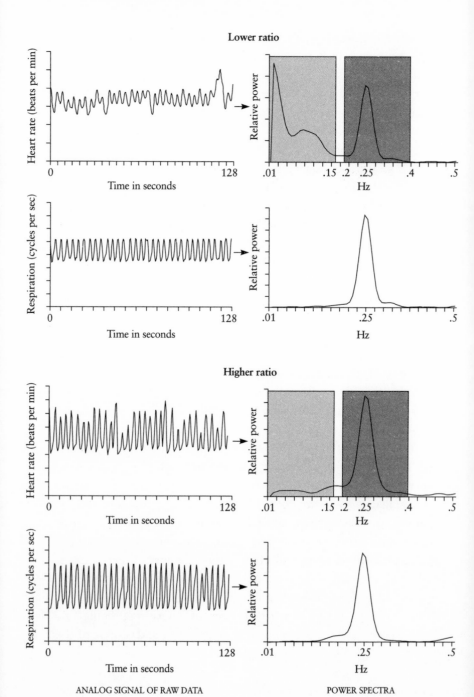

Lower ratio

Heart rate (beats per min)

Time in seconds

0   128

Relative power

.01   .15 .2   .25   .4   .5

Hz

Respiration (cycles per sec)

Time in seconds

0   128

Relative power

.01   .25   .5

Hz

Higher ratio

Heart rate (beats per min)

Time in seconds

0   128

Relative power

.01   .15 .2   .25   .4   .5

Hz

Respiration (cycles per sec)

Time in seconds

0   128

Relative power

.01   .25   .5

Hz

ANALOG SIGNAL OF RAW DATA

POWER SPECTRA

skin (Kistler, Mariauzouls, and von Berlepsch, 1998). However, some investigators believe that the skin is served by functionally distinct, sympathetically mediated vasodilation fibers whose activation produces surface warming (Loewy, 1990a,b). Activation of cholinergic-sympathetic fibers serving the sweat glands is often accompanied by dilation of surface capillaries and a warmer skin temperature (Cohen and Coffman, 1991). Thus, a psychological challenge could produce either a cooler or a warmer skin temperature, depending on which of the above systems was dominant.

We measured the temperature of the fingertips of the index, middle, and ring fingers of both hands with a thermography camera when the child was listening to a series of digits she had to remember (10 series of numbers were presented). The variables of interest were the initial temperature of the fingertips, the mean temperature over the whole series of memory items, and the magnitude of change in skin temperature over the 10 memory problems. This procedure was administered toward the end of the battery after the children had been in the laboratory for over 3 hours. We also measured the temperature of the ears, using a pediatric thermometer.

Because an increase in amygdalar activity should result in greater sympathetic reactivity, signs of sympathetic arousal should be more obvious in children who had been high-reactive infants, while high vagal tone should characterize low-reactives (Snidman et al., 1995). Earlier evaluations of these children revealed greater sympathetic activity in the cardiovascular system of high- compared with low-reactives. We knew from other studies that high vagal tone was associated with less irritability in 3-month-old infants (Huffman et al., 1994), easier socialization in 3-year-olds

4.7 Illustration of spectral analysis of a child with a low ratio of high to low power (above) and a high ratio of high to low power (facing page).

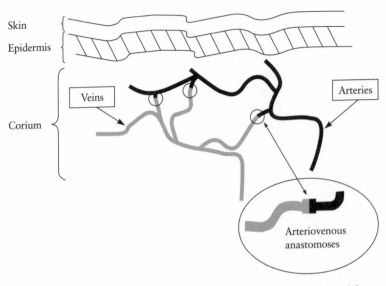

4.8 Schematic illustration of arteriovenous anastomoses. (Adapted from Wilger and Kaplan, 1984.)

(Porges et al., 1994), and greater sociability in 5-year-olds (Doussard-Roosevelt, Montgomery, and Porges, 2003).

Findings

The high/low ratio from the spectral analysis of heart rate best discriminated the high- from the low-reactives (see Table 4.8). The high-reactives had lower ratios ($F\ 2/141 = 2.43$, $p = .09$) and, by implication, a greater sympathetic and smaller vagal contribution to a resting supine heart rate. A post-hoc comparison revealed a significant difference between high- and low-reactives, with low-reactive boys showing the highest ratios and therefore the highest vagal tone.

The combination of the ratio and resting heart rate separated high- from low-reactives more effectively than either measure alone. Thirty-four percent of low-reactives, but only 16 percent

Table 4.8. Mean values for sympathetic variables for temperament and gender groups.

Variable	Low-reactives		Others		High-reactives	
	Boys	Girls	Boys	Girls	Boys	Girls
Heart period, eyes open	813	788	796	746	785	782
Heart period, speech	805	783	784	727	778	766
Ear temperature	96.6	96.8	96.6	97.0	96.5	96.5
Systolic, sitting	111	112	112	113	112	113
Systolic, standing	114	113	115	113	114	115
Diastolic, sitting	58	57	57	56	59	59
Diastolic, standing	60	60	61	62	63	61
Heart period, auditory test, phases 1–3	741	733	756	670	739	715
Heart period, auditory test, phases 4–6	735	710	727	665	710	710
High-low ratio spectral analysis	2.46	1.53	2.39	1.25	1.59	1.10
Index finger temperature, phase 1	30.9	30.4	31.8	31.4	32.3	31.3
Index finger temperature, phase 4	31.1	30.4	31.8	30.9	32.5	31.2

of high-reactives, combined a high ratio with a low resting heart rate (based on a z score of ±.00). However, 49 percent of high-reactives but only 37 percent of low-reactives showed the opposite pattern of greater power in the lower-frequency band and a higher resting heart rate (chi-square (1) = 4.9, $p < .05$). About 1 of 3 low-reactive boys had high vagal tone (high ratio combined with a low heart rate), compared with 1 of 6 high-reactives. One of every two high-reactives had high sympathetic tone (a low ratio with a high heart rate) compared with one of 3 low-reactives (see Figure 4.9). In addition, the boys and girls with higher sympathetic tone, compared with those with high vagal tone, had higher beta power at rest ($F\ 3/86 = 4.37$, $p < .05$) and warmer finger temperatures ($F\ 3/78 = 2.53$, $p = 06$).

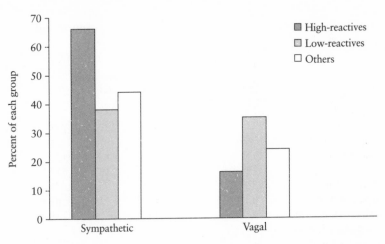

4.9　Percent of each temperamental group classified as sympathetic or vagal.

Relation to behavior. We compared boys and girls with high vagal tone with those who had high sympathetic tone, independent of their temperament. The latter group spoke less often, were more inhibited with the examiner, and were described by their mothers as relatively shy compared with the children with high vagal tone (F 3/86 = 3.54, p < .05 for the rating of inhibition; F 3/86 = 2.68, p = .05 for comments; F 1/111 = 5.35, p < .05 for the description of shyness; see Stifter and Jain, 1996).

The children with higher vagal tone also smiled more often during the second year (chi-square (1) = 3.8, p .05) and described themselves at age 11 as "happy" (they placed the Q-sort item, "Most of the time, I'm happy" in ranks 1–3). This difference was clearest among low-reactive boys. The low-reactive boys who were spontaneous with the examiner, described as sociable, and reported liking new activities had the lowest heart rates of any group. One of every 4 low-reactive boys, but only two high-reactive boys, had extremely high vagal tone and were described as minimally shy (z < −.5 for both variables). One of every 5 high-reactive boys, but only 1 low-reactive boy, had equally extreme values for sympathetic tone and a maternal description of shyness ($p < .01$ by the Exact Test).

Comparable differences emerged when the children were divided on the basis of behavior rather than on cardiac measures. The children who were spontaneous with the examiner ($z > .00$ for both comments and smiles) were more often vagal; those who were subdued were more likely to be sympathetic (chi-square (1) = 10.3, $p < .01$).

There was a linear increase, among girls, between number of comments with the examiner and resting heart rate; that is, girls became quieter as their heart rates rose, independent of their temperament ($F\ 2/105 = 4.5$, $p < .01$). Among boys, however, baseline heart rate interacted with temperament. Although low-reactive boys, like most girls, became quiet as their heart rates increased, the high-reactive and other boys spoke more often as their heart rates rose ($F\ 4/102 = 2.53$, $p < .05$). High-reactive boys with high heart rates were garrulous; high-reactive girls with equally high heart rates were very quiet.

Another example of the principle that the relations among variables can vary with temperament or gender is the fact that low-reactives with a high resting heart rate and high beta power were subdued, but high-reactives with the same heart rate and beta power were spontaneous. These data suggest that investigators should not treat shyness-sociability as a continuous construct. Like hyper- and hypothermia, the conditions that produce extreme shyness are not the opposite of those that produce a sociable persona.

Heart rate during test for auditory acuity. The test for auditory acuity, which occurred early in the battery, required a high level of motivation from most children. Although there was no significant difference between high- and low-reactives in mean heart rate during the first few trials of the lengthy procedure, more high- than low-reactives showed a linear increase in heart rate across the 15-minute test, and therefore ended up with the highest heart rates during the final trials of the 6-phase procedure. Sixty-one percent of high-reactive boys, but only 41 percent of low-reactive boys, attained their highest heart rate during the last two phases of the procedure (phases 5 or 6); 3 times as many low as high-reactives attained their highest heart rate early (during phase 1 or 2) (chi-square (1) = 4.8, $p < .05$). High-reactive boys had larger heart rate accelerations from phase 1 to phase 6 than low-reactive boys (chi-square (1) = 4.9, $p < .05$). These high-reactive boys with large accelerations spoke less frequently with the examiner, but this relation was absent for girls because many attained their highest heart rate early in the procedure and therefore could not show a large acceleration over the course of the test.

Heart rate acceleration to speech preparation. Many children showed a heart rate acceleration during the 1-minute interval when they were preparing a short speech they thought they would have to recite. Magnitude of heart rate acceleration to the speech challenge was correlated with the loss of alpha power (from eyes-open to speech) in girls, but with a gain in beta power in boys.

Because the combination of a large change in alpha or beta power and a large heart rate acceleration to the challenge might reflect a heightened state of sympathetic/cortical arousal, we created two contrasting groups. Group 1 ($N = 25$) contained boys who combined a large gain in beta power with a large heart rate acceleration and girls who combined a large loss in alpha power with a large acceleration to the challenge of preparing a speech. The contrasting group of 59 children consisted of boys with a small gain in beta power and girls with a small loss in alpha power, all of whom showed small heart rate accelerations. The former aroused group (which contained both high- and low-reactives) was rated as more inhibited with the examiner ($F\ 1/82 = 4.14$, $p < .05$).

Finger temperature. A higher resting heart rate was correlated with warmer finger temperatures across all children ($r = .3$, $p < .01$), implying greater blood flow to the periphery, due in part to sympathetic activity. More high- than low-reactives combined a high resting heart rate with warm fingers. The warmer finger temperatures among high-reactives could be due either to cholinergic fibers serving the sweat glands or sympathetically driven vasodilation.

The high-reactives, especially the boys, had warmer finger temperatures than low-reactives on all 3 fingers across all the digit memory trials (F 2/213 = 2.96, p < .05). Further, the finger temperatures of high-reactive boys, which were warm initially, became warmer over the trials; while the finger temperatures of the low-reactives, which were warm initially, were more likely to cool over the trials (chi-square (1) = 5.1, p < .05). High-reactive girls who began the procedure with warm fingers also showed cooling, but low-reactive girls whose fingers were cool initially continued to cool over the procedure. Thus, high-reactive boys were unique; they began the procedure with warm fingers and showed increased warming as the task became increasingly difficult.

Relation of finger temperature to behavior. The children who showed the greatest cooling over the digit memory items, due to vasoconstriction, had the lowest number of comments and smiles and were rated as inhibited with the examiner. In addition, high-reactives who cooled the most had higher fear scores in the second year (t (58) = 2.32, p < .05), were described by their mother as shy (t (48) = 2.88, p < .05), and reported "not liking new activities" (t (48) = 2.51, p < .05) compared with other high-reactives. Thus sympathetically mediated constriction of the anastomoses to a cognitive challenge was associated with signs of inhibition across all children.

It appears that two psychological states affected skin temperature. The state created by the prior 3 hours of testing was associated with a warmer skin temperature, especially among high-reactives. The more acute state created by the challenge of committing strings of numbers to memory was associated with constriction of the anastomoses, a cooler skin temperature, and subdued behavior.

Combining ear and finger temperatures. The orienting reaction to a stimulus is characterized by dilation of capillaries in the head (indexed in part by a warmer ear temperature) and constriction at the periphery (indexed in part by cooler finger temperatures). We divided the temperature distributions for both index fingers and both ears into terciles to determine which children showed the orienting pattern and which children displayed either warm ears combined with warm index fingers or cool ears with cool index fingers. The low-reactive boys were more likely than high-reactive boys to show the orienting pattern (29 versus 8 percent), suggesting greater engagement in the laboratory procedures. The boys who clearly endorsed the item "Most of the time, I'm happy" (they placed this item in rank 1) most often showed the orienting pattern. By contrast, more high than low-reactive boys combined warm index fingers with warm ears (41 versus 18 percent; chi-square (1) = 7.0, $p < .01$), and more high- than low-reactive girls combined warm fingers with a cool ear temperature (55 versus 2 percent; chi-square (1) = 16.6, $p < .01$).

Summary

The combination of the spectral analysis and resting heart rate best differentiated high- from low-reactives, and the measures of sympathetic tone in the cardiovascular system were the better correlates of social behavior with strangers than the EEG or Wave 5 variables. The mean z score for (1) resting heart rate, (2) ratio from the spectral analysis, and (3) temperature of the index fingers was positively correlated with the rating of inhibited behavior across all children ($r = .3$, $p < .01$). It is not clear why the cardiovascular measures were more closely linked to social behavior than the other biological variables. One possibility is that activity in autonomic targets is more directly associated with a child's conscious feeling tone than the EEG or brain stem measures. Afferent information from autonomic targets ascends first to the medulla, the amygdala, and finally to the orbitofrontal prefrontal cortex. Activation of the latter site contributes to a child's feeling of tension or uncertainty and could create a state that mediated a subdued posture with the examiner.

EVENT-RELATED POTENTIALS

Rationale

The event-related potential (ERP) is a time-locked, post-synaptic potential generated by large numbers of cortical neurons to a specific stimulus. Because the voltages generated by a single stimulus are small and variable, it is necessary to average the potentials over many stimulus presentations. The synchronized neural activity creates a dipole, defined as a separation of charge in a volume conductor, which occurs when an excitatory or inhibitory input from a neighboring neuron alters the charge on a post-synaptic cell. If the input on the neuron is excitatory, positively charged ions flow into the cell to create a negative charge in the contiguous extracellular space, called the sink. These positive ions flow through part of the neuron and then exit back into the extracellular space, to create a positively charged area called the source (see Figure 4.10). If the synapse is inhibitory, the sink will have a positive charge and the source a negative one. The disparity in charge between the sink and the source establishes a dipole.

When a negative source is located closer to the scalp than a positive sink, a negative wave form is measured from the scalp electrodes. When a positive sink is located closer to the scalp then a negative source, a positive wave form is detected. The polarity of each wave form, whether negative or positive, is indicated by an *N* or *P* followed by a number that indicates the usual latency to the peak magnitude following the onset of a stimulus. If the sink and the source are distributed radially, the dipoles cancel each other and there will be no measurable wave form. However, if the dipoles of the sink and the source are aligned in a non-radial fashion,

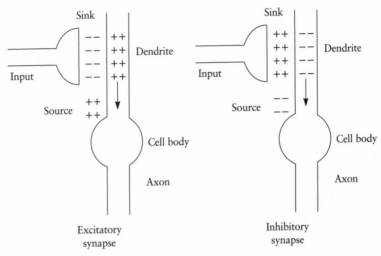

4.10 Schematic illustration of the source and the sink for excitatory and inhibitory inputs to a neuron.

and the dendritic trees are oriented in the same direction and the axons in the opposite direction, the synchronized activity will create a measurable dipole (Nelson and Monk, 2001).

Theoretically, there is an infinite number of dipole distributions that could produce a specific ERP wave form. As a result, investigators cannot infer with certainty the source of a particular wave form at a particular scalp site. It is believed that the wave forms are due primarily to neuronal activity in the first 4 layers of the cortex, suggesting that event-related potentials provide more information on sensory input and interactions across the cortex than on efferent connections between the cortex and subcortical sites.

The wave forms that appear during the first 100 msec represent neuronal reactions to the sensory quality of an external stimulus and are called *exogenous.* The wave forms generated after 100 msec, called *endogenous,* reflect a comparison with a preceding event and/or the activation of a schema or a semantic representation acquired in the past that awards meaning to the event (Van

Rollen and Thorpe, 2001). The wave form to faces compared to objects was different as early as 110 msec after the stimulus appeared, and a negative wave form at 170 msec occurred only to photos of faces and not to varied examples of 7 different classes of objects (Itier and Taylor, 2004).

The first wave form that represents the detection of a sensory event that deviates from the immediate past is called N2 because it has a negative polarity and usually peaks at about 200 msec in adults. When the discrepant event is a change in the frequency, loudness, or duration of a tone, this wave form is called mismatch negativity (MMN) (Naatanen and Winkler, 1999).

Two succeeding wave forms, called P3 and N4 in adults, appear between 250 and 800 msec. The wave form called the novelty P3 (P3a) has a typical latency to peak voltage of 300 msec, is usually larger at frontal than parietal sites, and is prominent when the subject is attentive but has no cognitive task to perform (Geisler and Murphy, 2000). If a subject is given the assignment of responding to an infrequent target, the wave form is called the P3b and the largest magnitudes appear at parietal rather than frontal sites. Most scientists believe that the novelty P3 and the P3b should be regarded as separate phenomena (Goldstein, Spencer, and Donchin, 2002).

The wave form called the N4 usually appears between 350 and 800 msec with a mean latency of 400 msec. If a printed word at the end of a sentence is semantically inconsistent with the meaning of the prior words (for example, "Carrots are good things to breed"), the N4 occurs at about 400 msec. Federmeier and Kutas (2002) presented adults with pairs of sentences, but the last idea in the second sentence of each pair was displayed as a picture rather than as a word. Some final pictures were congruent with the meaning of both sentences, while some were not. There were two types of discrepant violations. The less serious discrepancy consisted of a picture that belonged to an appropriate semantic category but was not the best fit for that pair of sentences. The most serious discrepancy was a picture that did not fit the meaning of the sentences. For example, if the semantically most congruent picture was a poodle, a mild discrepancy was a picture of a Dalmatian. A more serious discrepancy would be a picture of a donkey. The more serious discrepancies produced larger N4 wave forms, implying that the most discrepant violations of meaning produced the larger N4 magnitudes.

Both the P3a and N4 reflect the brain's detection of an event that is unexpected or that transforms a familiar representation. A letter string that is possible in English but is not a legitimate word (such as *tolip*) elicits an N4 (Deacon et al., 2004). The magnitudes of both wave forms show their largest increase between 3 and 5 years of age, when children begin to show reliable integration of past and present. The novelty P3 and N4 can occur to discrepant or unfamiliar events in any modality, and more surprising discrepant events elicit larger magnitudes (Olivares, Iglesias, and Rodriguez-Holguin, 2003).

One hypothesis that might render the varied wave forms comprehensible assumes that the usual latency of the wave form is an index of the amount of "work" the brain must perform on the stimulus. If the brain only has to detect the fact that a tone is louder than a prior series of identical tones that occurred several seconds earlier, the usual latency is about 200 msec—this is the N2. If the brain must relate the stimulus to schemata acquired in the distant past, the latency is about 100 msec longer—this is the P3a. And if the brain must relate the event to semantic, or semantic combined with schematic, representations, the usual latency is about 400 msec—this is the N4. Whether the wave form is positive or negative often depends on whether the person can or cannot categorize the event—the P3 is more likely for the former; the N_4 for the latter.

The rationale for recording ERP wave forms in our 11-year-old sample was based on the assumption that the threshold of reactivity in the amygdala in response to discrepant events, especially in the lateral nucleus, is lower in high- than in low-reactive infants. Direct recordings from the amygdalar neurons of monkeys revealed a group of cells in the basolateral nucleus that respond to novel events (Wilson and Rolls, 1993; Onu and Nishijo, 2000). The largest N4 magnitudes often appear in sites close to the amygdala (Halgren et al., 2002). Further, when adults passively viewed pictures of modeled objects or scenes, without any task to perform, fMRI data revealed activation of areas contiguous to the amygdala (the hippocampus and inferior temporal cortex; Jessen et al., 2002). In addition, adults showed greater amygdalar activity to unfamiliar than to familiar faces with neutral expressions, and adults who had been inhibited as children had the greatest activity (Schwartz et al., 2003a). Veterans with PTSD showed larger P3 wave forms in response to novel sounds than combat soldiers without PTSD (Kimble et al., 2000); and the magnitude of the P3 to a discrepant tone or to novel words was larger in panic disorder patients than controls (Clarke et al., 1996; Windmann, Sakhaut, and Kutas, 2002). By contrast, adolescents with conduct disorder had smaller P3 wave forms (Kim, Kim, and Kwon, 2001).

We predicted a larger ERP wave form in response to unfamiliar scenes among high- than among low-reactive children for several reasons. First, projections from the lateral nucleus of the amygdala to the locus ceruleus and ventral tegmentum should be accompanied by increased levels of norepinephrine and dopamine in the cortex. Second, projections from the amygdala to the cortex should enhance the synchronization of cortical pyramidal neurons. Third, the central nucleus of the amygdala sends projections to the dorsal nucleus of the lateral geniculate of the thalamus, which in turn projects to cortical neurons (Cain, Kapp, and Puryear, 2002). Each of these mechanisms could theoretically lead to larger ERP wave forms in response to discrepant events (Aston-Jones and Bloom, 1981).

Previous research indicated that infants and preadolescent children usually show a negative rather than a positive wave form in response to discrepant visual events, even though adults can show a negative wave form to unfamiliar events (Ganis and Kutas, 2003).

The negative wave form in children, called Nc, was the variable of interest in our analyses. There is some disagreement over whether the Nc, which has a latency of about 600 msec in infants, 400 msec in preadolescent children, and 200 msec in adults, should be viewed as homologous to the N2, N4, or the P3a in adults. Courchesne (1978) believes that the Nc in children should be treated as similar to the adult N2. But Ceponiene et al. (2004) regard this wave form as distinct. Investigators presented 10- to 13-year-olds with 3 types of auditory events: (1) a syllable that occurred 80 percent of the time, (2) a different syllable that occurred 10 percent of the time, and (3) a series of different novel sounds for the remaining 10 percent of the trials (clicks, chirps, and so on). The fact that the mismatch negativity wave form was distinguished from the Nc—the latter occurred between 450 and 1,100 msec—supports the view that the Nc is a reaction to a discrepancy that requires more cognitive work. Thus, we assumed that the Nc in our subjects represented the brain's reaction to a discrepant event that required some cognitive work (Johnstone and Barry, 1999; De France et al., 1997; Nelson et al., 1998).

Each child was presented, through goggles, with two series of chromatic pictures, with 169 pictures in each series. The two series were always presented in the same order. In the first series, 70 percent of the pictures were of the same item (a fireworks display), 15 percent (25 pictures) were of the same flower (called the oddball stimulus), and the remaining 15 percent were each different but ecologically valid (a chair, fork)—these pictures were called novel valid. In the second series, the frequent picture, presented 70 percent of the time, was a yellow fire hydrant, the oddball stimulus was of a very different flower, and the remaining 15 percent were each different but ecologically invalid (a chair with 3 legs, a baby's head on an adult body)—these pictures were called novel invalid.

The child had no task assignment; he was simply asked to remain still without blinking and to look at each picture. Each picture was shown for 1 second with an inter-stimulus interval of 1.2 seconds. An important reason for not requiring the child to do anything but simply sit quietly and look at the pictures is that fMRI measures of brain activity reveal that when subjects are required to attend to a stimulus and to make a finger movement when a target appears, unique patterns of activation occur that are not observed when the subject has no task assignment (Indovina and Senes, 2001).

ERPs were recorded from F3, Fz, F4, P3, Pz, and P4, referenced to Cz for recording and re-referenced to average linked earlobes offline. Each scalp site was prepared by gently abrading with a conductive abrasive and attaching a silver electrode using an electrolytic cream. Vertical eye movements were recorded from Ag/AgCl electrodes placed supra- and suborbitally to the left eye. Horizontal eye movements were recorded from Ag/AgCl sensors placed at the outer canthus of each eye. Data were sampled at 1,000 Hz, band-pass filtered at .08–100 Hz, resampled at 250 Hz, and digitally filtered at .08–25 Hz. Eye movement was corrected offline.

Data were collected for a 2.2-second recording epoch which included a 100 msec pre-stimulus baseline and a 1-second presentation of the picture. The individual ERP wave forms for each trial were combined to produce average wave forms for each category of stimulus (frequent, oddball, novel valid, and novel invalid). An occasional off-scale trial was rejected, but the mean value for each category of stimulus contained a minimum of 18 trials. If the number of trials was below 18 for any class of picture, that child was removed from the analysis (only two children were removed). The integrated voltages for the Nc wave form were calculated using a computer algorithm.

A visual inspection of the grand averages and a principal components analysis, using all leads and picture types, identified two components. One variable was the total amount of negative voltage integrated from 80 to 400 msec; the second area was the integrated voltage from 400 to 1,000 msec. The two variables were computed separately for each class of picture at frontal, central, and parietal sites. The decision to concentrate the analysis on the integrated voltages from 400 to 1,000 msec is supported by empirical evidence. For example, it was the ERP wave form after 400 msec that differentiated subjects who received an unexpected painful stimulus to the skin from those who expected the pain (Legrain et al., 2003). Second, semantically inconsistent words at the end of a sentence (such as "The tree had four sockets") do not produce a broadly distributed wave form until 400 msec (Halgren et al., 2002). These data imply that the wave forms after 400 msec reflect more elaborate cognitive activity and therefore might provide better differentiation between high- and low-reactives.

In addition, 35 minutes after the end of the second series, children were given a recognition memory test consisting of 48 pictures—24 from the novel valid or novel invalid series that had been seen, and 24 pictures not seen before. The children did not know that this test would be administered.

Findings

A clear negative wave form appeared at 150 msec, with peak voltage at about 400 msec and a return to baseline by 1,000 msec (see Figure 4.11). The values were largest at frontal and central sites and much smaller at parietal sites (see Table 4.9). The integrated voltages from 400 to 1,000 msec were larger than those from 150 to 400 msec, and the former variable, which we call Nc, was the one we explored in detail.

The magnitude of Nc was a function of degree of discrepancy. The Nc values were largest to the novel valid and invalid scenes, next largest to the two oddball stimuli, and always smallest to the frequent pictures (see Figure 4.12). The effect of picture category

Variable	Low-reactives		Others		High-reactives	
	Boys	Girls	Boys	Girls	Boys	Girls
Fz						
Frequent 1	3248	2966	3019	2612	3484	3200
Oddball 1	3856	4078	3708	3596	5437	4276
Novel valid	7959	7515	7855	6997	7579	8320
Novel invalid	8865	7909	9388	7604	8474	8861
Frequent 2	4021	3823	3631	3627	4201	4305
Oddball 2	4819	4501	4757	3232	3931	4753
Cz						
Frequent 1	2916	1953	1829	1932	2264	2116
Oddball 1	2594	2494	2805	2981	3302	2914
Novel valid	4749	4663	5659	4714	4976	5349
Novel invalid	5000	5161	6152	5151	5921	6781
Frequent 2	2535	2512	2487	2653	2614	3164
Oddball 2	2993	3366	3343	2193	2650	3468

was highly significant across frontal and central sites (F 11/2211 = 161.3, p < .001). In addition, we ranked each child's Nc values—oddball, novel valid, and novel invalid—at Fz and Cz sites. A majority showed larger values to novel invalid than to novel valid, larger to novel valid than to oddball, and larger at Fz than at Cz. Because the magnitude of Nc was correlated with the degree of discrepancy, we could ask whether the temperamental groups differed on this measure.

The high-reactives had significantly larger values to the oddball stimulus in the first series at Fz and to the novel invalid pictures in the second series at Cz (t = 2.19, p <.05 for oddball; t = 1.74, p = .08 for the novel invalid). Forty-three percent of high-reactives, but only 25 percent of low-reactives, had high values (above the median) to both the first oddball at Fz and the novel invalid pictures at Cz (chi-square (1) = 6.5, p < .05). This was particularly

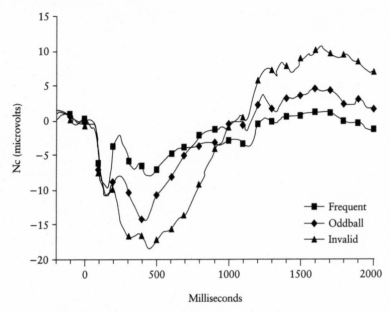

4.11 Illustration of the Nc wave form at frontal sites to the frequent, oddball, and invalid pictures.

true for high-reactives who showed right frontal activation and high beta power under the eyes-open condition.

We also regressed the Nc values to the oddball flower in the first series on the values for the frequent picture in the first series, and the values for the novel invalid scenes on the values for the frequent stimulus in the second series, and examined the residuals—that is, the degree to which a child had larger values than predicted by his reaction to the frequent stimulus in that set. More high-than low-reactives had positive residuals for both variables; more low-reactives had negative residuals (chi-square (1) = 6.4, $p < .01$, for oddball; chi-square (1) = 6.1, $p < .01$ for the novel invalid pictures). A similar difference emerged when we subtracted the value for the oddball flower in series 1 from the mean value for the first frequent picture, and subtracted the value for novel valid from the value for novel invalid. Sixty-four percent of high-reactives but

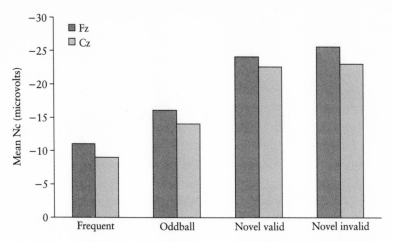

4.12 Mean magnitude of the Nc wave form to 4 categories of stimuli across all children.

only 46 percent of low-reactives had larger values to novel invalid than to the novel valid scenes (chi-square (1) = 4.2, p <.05); these children tended to be the more spontaneous children.

We standardized the distributions of two difference scores: (1) oddball 1 minus frequent 1 at Fz, and (2) novel invalid minus novel valid at Cz (see Figure 4.13). As expected, 56 percent of high-reactives but only 38 percent of low-reactives had z scores greater than .00 on both variables (chi-square (1) = 4.3, p <.05).

The correlations between the Nc values to the oddball and novel pictures at frontal and central sites were equally large for all 3 temperamental groups (r = .8). However, the correlations between values at frontal or central sites on the one hand and parietal sites on the other were only high among high-reactives (r = .4 for frontal with parietal; r = .7 for central with parietal; see Table 4.10). The comparable correlations for the other two temperaments were .3 and .1. That is, only high-reactives showed coherent Nc values across disparate cortical sites (chi-square = 4.8, p < .05), implying the recruitment of neuronal activity over a broader cortical area.

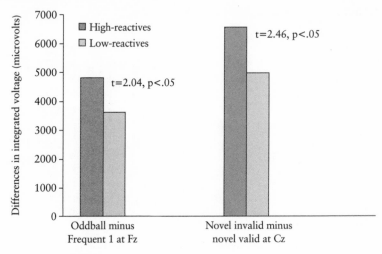

4.13 Oddball minus frequent 1 at Fz and novel invalid minus novel valid at Cz for high- and low-reactives.

Because amygdalar excitability should enhance both Wave 5 and Nc to the novel invalid scenes, we compared high- and low-reactives who were high or low on both variables (based on a *z* ±.00 to Wave 5 and novel invalid minus novel valid). More high- than low-reactives had high values on both measures. Further, the high-reactives who had high values for both Wave 5 and Nc displayed the fewest number of smiles. Low-reactives with high values on both variables smiled more often (78 percent of the high-reactives had z values < .00 versus 46 percent of the low-reactives, chi-square (1) = 3.6, *p* < .05). Once again, the indirect biological signs of amygdalar excitability were associated with a serious mood in high-reactives but a more sanguine mood in low-reactives.

Asymmetry of Nc at frontal sites. There was substantial variation in the degree to which the frontal Nc values to the discrepant pictures were larger on the left (F3) or the right (F4). Most children

Table 4.10. Correlations among different sites for 4 classes of stimuli.

	Oddball 1	Oddball 2	Novel valid	Novel invalid
Fz with Cz				
Low-reactives	.71	.87	.81	.81
Others	.75	.81	.76	.84
High-reactives	.70	.75	.75	.81
Cz with Pz				
Low-reactives	.31	.33	.18	.38
Others	.55	.27	.46	.58
High-reactives	.71	.66	.69	.68
Fz with Pz				
Low-reactives	.03	.17	−.01	.18
Others	.21	.09	.20	.39
High-reactives	.42	.37	.46	.56

had larger values on the left; a majority of these children were relaxed and spontaneous. Children with larger values on the right were more subdued. If implicit verbal labeling of the pictures were associated with greater activity on the left, while encoding pictures with schemata were preferentially linked with greater activity on the right, perhaps the sociable children activated semantic networks, while the more subdued children, usually high-reactives, preferentially attended to the perceptual features of the pictures. The profile of brain activity to a stimulus, as measured by PET, depends on the cognitive operations the subject performs on a particular stimulus (Fias et al., 2002).

The accuracy of recognition memory for the invalid and valid pictures assessed a half hour later was related to larger Nc values on the left side but not to mean Nc values. Three-fourths of the children with the best memory scores (47 or 48 correct out of a total of 48, and comprising 48 percent of the sample from all temperament groups) had larger values on the left than on the right. We suspect that these children implicitly named the pictures, and as a result had a better memory score and larger Nc values over the left hemisphere.

An alternative interpretation assumes that the discrepant scenes created more uncertainty in the subdued children and, as a result, neuronal activity was greater on the right. This interpretation finds support in the fact that panic patients, compared with controls, have larger ERP wave forms to novel words in the right frontal area (Windmann, Sakhaut, and Kutas, 2002).

Because the magnitude of the left minus right difference was similar (within a child) for the novel valid and invalid pictures ($r = 0.5$), we computed a mean left-right difference at frontal sites across both classes of pictures and divided that distribution into terciles. The children in the lowest tercile, across all temperamental groups, had larger Nc values on the right side; those in the highest tercile had larger Nc values on the left side. The former group had larger Nc values to the novel invalid pictures ($F\ 2/204 = 3.18, p < .05$ for novel invalid). The larger the voltage difference between right and left, favoring the former, the larger the Nc to the invalid scenes, and the more likely the child displayed a large Wave 5 (chi-square $(1) = 6.6, p < .01$). This effect was similar for all 3 temperamental groups.

Summary

The quintessential difference between high- and low-reactives is the differential reactivity to unfamiliar events, which we suppose is due to variation in amygdalar activity. This evidence implies that the two temperamental groups preserved this biological property for more than 10 years.

STARTLE AND CORRUGATOR ACTIVITY
Rationale

The startle reflex has a rich history. Although a body startle in response to a sudden loud sound or unexpected touch is common, this reflex became a focus of interest about 50 years ago when, as noted in Chapter 3, psychologists at the University of Iowa decided to use the conditioned startle reflex as an index of fear in rats (Brown, Kalish, and Farber, 1951). After pairing a neutral stimulus (a combination of light and buzzer) with electric shock, they observed an enhanced body startle to a loud acoustic probe when the conditioned stimulus preceded the probe. They concluded that this larger startle—called potentiated startle—reflected fear, on the assumption an animal anticipating shock should be afraid. The startle was only potentiated to a conditioned stimulus that signaled electric shock; it was not larger when the animal was given an electric shock in one context and administered the acoustic probe in a very different place. That is, the brain state created by the tingling sensation of electric shock does not produce a potentiated startle.

This paper became popular because the concept of fear was central to both behaviorism and psychoanalytic theory. A few decades later, behavioral biologists began to study the brain bases for the potentiated startle. Davis and colleagues, who probed the original Iowa discovery with the elegant tools of the neuroscientist, learned that the amygdala primes a brain stem nucleus (nucleus reticularis pontis caudalis) that mediates the startle reflex (Davis, 1994; Lee et al., 1996; Walker and Davis, 1997). Therefore, an excited amygdala should be accompanied by a potentiated startle. Because an intact amygdala is necessary for the acquisition of a number of conditioned responses presumed to reflect a fear state (for example, bodily freezing), these discoveries implied that potentiated startle might be a sensitive measure of fear in human subjects.

If this hypothesis were valid, psychologists would possess a simple way to quantify an elusive but theoretically important family of psychological states. However, the ethical restraints imposed by university review committees prevent scientists from using electric shock with human subjects, most of whom would refuse to participate in an experiment that required them to display full body startles. It was necessary to find a substitute for the unconditioned response of the full body startle, as well as a replacement for the electric shock.

Psychologists chose the reflex eyeblink in response to an acoustic probe (a brief burst of loud white noise) as a substitute for the body startle because the blink always accompanies the latter reaction. These scientists ignored the fact that different neural circuits mediate the body startle and the eyeblink (Holstege, Van Ham, and Tan, 1986), as well as the fact that the eyeblink reflex is potentiated when subjects, who are in a relaxed emotional state, contract their facial muscles voluntarily (Aniss, Sachdev, and Chee, 1998).

This last fact is important because the blink can be enhanced by excitability in the facial nerve, independent of activity in the amygdala or brain stem (Miwa et al., 1998). The excitability of the motor neurons of the facial nucleus, which innervate the orbicularis oculi muscles that produce the blink, are influenced by many brain structures, including the red nucleus, olivary pretectal nucleus, pontine tegmentum, and spinal trigeminal nucleus (Esteban, 1999; Goodmurphy and Wik, 1999). Anxious adults have greater muscle tension in many body sites, not just the orbicularis oculi (Fridlund et al., 1986). Further, neurons in the deep superior colliculus and the mesencephalic reticular formation (in rats) influence the magnitude of startle to a visual conditioned stimulus. The injection of a GABA agonist into these areas reduces the magnitude of the potentiated startle, without affecting the magnitude of baseline startle (Meloni and Davis, 1999). Finally, sympathetic activation, which accompanies a state of uncertainty, can reduce startle magnitudes (Tavernov et al., 2000).

Although psychologists tried various substitutes for a conditioned light or tone paired with shock, eventually pictures judged by observers to be unpleasant or aversive became the usual way to potentiate the eyeblink reflex. The rationale for this decision rested on the premise that pictures of bloodied soldiers, venomous snakes, and gun barrels provoke a brain state in humans that is essentially similar to the one created by presenting rats with a light or tone that had been associated with shock. These scientists assumed that unpleasant or aversive pictures had become conditioned stimuli linked to actual events in the subject's past that were painful or unpleasant (Bradley, Cuthbert, and Lang, 1996).

Although this argument is vulnerable, skepticism was muted when investigators found repeatedly that the magnitude of the blink reflex in response to an acoustic probe was usually larger when an adult was watching an aversive picture, compared with scenes that had been judged as symbolic of pleasant or neutral events (Vrana, Spence, and Lang, 1988; Bradley, Cuthbert, and Lang, 1999). The fact that incarcerated psychopathic subjects did not show a potentiated startle to unpleasant pictures, while control subjects did, was treated as support for the original assumption even though the concept of unpleasant events is too broad. Films that are symbolic of harm (for example, snakes and guns) produce different heart rate reactions than films depicting surgery, which often evoke feelings of disgust (Palomba et al., 2000). Hence, in a relatively short time, many psychologists accepted as a proven fact that a potentiated eyeblink reflex was a sensitive index of an aversive psychological state that might be related to states of fear or anxiety in humans. There are, however, some problems with this assumption.

First, although women have larger startles to unpleasant than to pleasant pictures (Lang, Bradley, and Cuthbert, 1992; Codispoti, Bradley, and Lang, 2001, Bradley and Lang, 2000), men often have larger startles than women to pleasant pictures, and they rate them as more arousing, in part, because the set of pleasant pictures used most often contains several erotic scenes. The unexpected appearance of a nude can be alerting and can potentiate startle (Schmidt, 2002). Similar gender differences occur in children (McManis et al., 2001).

One sample of men showed large startles when they first watched an erotic film, but both the magnitude of startle, as well as penile tumescence, decreased as the same film was shown repeatedly for a total of 18 times (Koukounas and Over, 2000). Further, many conditions that produce potentiated startle are not obviously aversive. The startle is potentiated, for example, when adults have simply been sitting for a short period in a dark laboratory room (Grillon et al., 1991) or have been exposed to an unpleasant odor (Ehrlichman et al., 1995; Miltner et al., 1994). It is not obvious that a dark room, which occurs every night before the pleasure of sleep, or an unpleasant odor, which is encountered frequently in large cities, creates a defensive state. It is more likely that these events in the laboratory setting are unexpected and provoke thought.

In adults, benzodiazepenes, which reduce feelings of anxiety as well as baseline startle magnitudes, have no effect on the magnitude of potentiated startle to the anticipation of electric shock. This fact is inconsistent with the claim that potentiated startle indexes a state of anxiety in humans (Baas et al., 2002). More important, experimental induction of an unpleasant feeling state does not always produce a potentiated startle. The emotional states of adults were manipulated by having one group of young men self-administer the drug prednisone—a corticosteroid—on 4 consecutive days, while a control group received a placebo. Although the former group reported more intense, unpleasant feelings and displayed an increase in right over left frontal activation in the EEG, they did not show larger startles than controls to unpleasant pictures (Schmidt et al., 1999). This result is congruent with the fact that 19 war veterans suffering from PTSD, compared with 74 veterans without PTSD, showed larger heart rate reactions to acoustic probes but did not show larger startles (Orr et al., 1997; Grillon et al., 1998; Miller and Greif, 2002, personal communication). And male college students belonging to 4 different groups—very high levels of anxious apprehension; anxious arousal; anhedonic depression; or none of these states (that is, they were neither anxious nor depressed)—showed equivalent startles to unpleasant pictures (Nitschke et al., 2002).

One day after rats were subjected to the "stress" of a 5-minute exposure to a cat, the animals showed potentiation of neural transmission to the amygdala and less exploration of an unfamiliar area, but did not show larger startles than controls (Adamec, Blundell, and Collins, 2001). This result is inconsistent with the hypothesis that magnitude of startle is a sensitive index of a state of defensiveness or fear mediated by the amygdala.

Perhaps the most important critique is that humans usually startle to a sudden loud sound when they are deeply engrossed in thought. For example, the blink is enhanced when subjects have to keep track of ellipses that remain on a screen for a longer than usual duration (Lipp, Siddle, and Dall, 1997; Lipp and Siddle, 1999; Filion, Dawson, and Schell, 1998; Lipp, Siddle, and Dall, 2000). Aversive pictures, a dark room, or an unpleasant odor in the laboratory might elicit a potentiated blink because each can provoke cognitive activity. When adults were asked to imagine past incidents that had a great deal of personal meaning, startle magnitudes were equally large to both pleasant and unpleasant memories (Miller, Patrick, and Levenson, 2002), leading the investigators to suggest that startle is potentiated when an elaborate associative network is engaged. Acoustic probes delivered while adults were working on cognitive problems (anagrams or arithmetic) produced the same magnitude of potentiated startle as unpleasant pictures (Sorenson, McManis, and Kagan, unpublished; Blumenthal, 2001).

The suggestion that startle is potentiated when a subject is involved in thought is supported by a study of adults who saw 4 types of pictures: physically unpleasant, socially unpleasant, pleasant, and neutral. The time of delivery of the acoustic probe varied from 300 msec to 4.5 sec. The physically aversive scenes only produced larger startles when the acoustic probe occurred later than 800 msec after the onset of the picture. Startles were also larger when the acoustic probe was presented 10 sec rather than 3 sec after an aversive picture had appeared (Sutton et al., 1997). Thus, individuals showed potentiated startle when they had time for the evocation of associations to unpleasant pictures. That is why the magnitude of startle to neutral and pleasant pictures increased linearly as probe time increased (Lethbridge et al., 2002). A state of cognitive engagement can also potentiate startle in children under 3 years of age; startles were larger when behavioral and heart rate reactions indicated attentional involvement with interesting, nonaversive stimuli (Richards, 2000).

It is relevant that men showed a decrease in heart rate rather than an increase to angry faces, implying that the former engaged their attention (Jonsson and Sonnby-Borgstrom, 2003). Chimpanzees, too, showed a decrease in heart rate to a photo of an unfamiliar chimp but a rise in heart rate to an animal that the chimpanzee recognized as very aggressive (Boysen and Bernston, 1989). One reason aversive pictures were better remembered after a long delay than neutral pictures is that they generated more elaborate cognitive processes (Ochsner, 2000).

On the other hand, the magnitude of the blink reflex was suppressed when subjects were attentive to a stimulus but did not elaborate it cognitively (Schmidtke and Buttner-Ennever, 1992). For example, adults were less likely to dwell on films depicting surgery than on films depicting violence and showed smaller startles to the former (Palomba et al., 2000; Kaviani et al., 1999).

Our laboratory found no relation in 6- to 8-year-olds between potentiated startle to aversive stimuli and contemporary shyness or fearful behavior during the second year (Kagan et al., 1999). Schmidt and colleagues (1999) also reported that 7-year-olds who were rated as very shy failed to display larger potentiated startles than non-shy children to the unpleasant demand to stand up and give a speech, even though this challenge produced right frontal activation in the EEG. Surprisingly, adolescents classified as uninhibited in the second year showed greater potentiated startle to aversive pictures than adolescents classified earlier as inhibited (Balaban, Snidman, and Kagan, unpublished).

The failure of formerly inhibited children to show larger potentiated startles to unpleasant pictures than uninhibited children might be due, in part, to activity of the inferior colliculus, the source of Wave 5 in the BAER. Projections from the inferior colliculus to the brain stem can suppress the magnitude of body startle in rats (Leitner and Cohen, 1985; Li, Prieber, and Yeomans, 1998). We have already noted that an activated amygdala, which is the presumed basis for potentiated startle, enhances the excitability of the inferior colliculus. Thus, an event that activated the amygdala might excite the inferior colliculus and, as a consequence, mute startle magnitude.

This review provides a reasonable basis for questioning the popular hypothesis that a potentiated eyeblink while exposed to an event that observers judge as symbolic of unpleasant or aversive events necessarily reflects an anxious or defensive emotional state in a subject, even though this assumption might be valid for some subjects on some occasions (Filion, Dawson, and Schell, 1998). People judge scenes as beautiful or ugly, and menu items as delectable or unappealing, but it is not obvious that these cognitive judgments generate a feeling that matches the semantic label. But there are good reasons to entertain an alternative notion—that the blink reflex is likely to be potentiated when humans are cognitively engaged.

We quantified the magnitude of the eyeblink startle and accompanying corrugator activity (the forehead muscles) to threatening and non-threatening incentives. The threatening incentive in the first procedure was a light that warned of a possible delivery of an air puff to the throat. The threatening event in the second procedure was 9 pictures of aversive scenes interspersed with 9 neutral and 9 positive pictures taken from the Lang series (Lang, Bradley, and Cuthbert, 1992). The acoustic probe producing the eyeblink startle was delivered for 500 msec at 90 dB.

In the first procedure, which involved the aversive air puff to the throat, the child first received 5 baseline acoustic probes and was then presented with 16 trials, 8 with a warning light on indicating an air puff might be delivered and 8 trials with no light on indicating that no air puff would be delivered. The two types of trials were administered on a random schedule.

In the second procedure, the child saw 27 pictures presented in a random order: 9 pleasant, 9 unpleasant, and 9 neutral, displayed through goggles. Each picture was shown for 6 seconds; inter-trial intervals varied from 1.2 to 2.4 seconds. The startle stimuli were 90 dB bursts (.5 seconds of white noise presented binaurally). The acoustic startle probes were delivered randomly between 3.5 and 5.5 seconds after the onset of the picture. Six startle probes were delivered between the pictures and 5 probes were delivered before the first picture was presented. Following this procedure the room was darkened and children received an additional set of acoustic probes while sitting quietly in the dark.

The eyeblink reflex was measured by recording electromyographic activity over the orbicularis oculi region under the left eye with Ag/AgCl surface electrodes. The raw EMG signal was amplified ($\times$ 10,000) and band-pass filtered between 10 Hz and 300 Hz, and digitally sampled at 1,000 Hz from 50 ms before the startle probe occurred until 250 msec after probe onset. Corrugator activity was recorded over the left corrugator muscle region with Ag/AgCl surface electrodes. The raw EMG signal was amplified ($\times$ 10,000) and band-pass filtered with a low cutoff of 10 Hz and a high cutoff of 300 Hz and digitally sampled at 1,000 Hz.

Startle responses were scored offline as the peak of the rectified orbicularis oculi response in microvolts for each trial. Startle magnitudes were averaged for each picture category to produce a mean startle amplitude for the pleasant, neutral, and unpleasant picture types. Corrugator EMG signal was digitally refiltered with band pass of 98 Hz to 250 Hz. The refiltered and rectified signal was averaged over the entire 6-second picture-viewing period, and we computed deviations from a baseline period of 500 msec to obtain a change score for corrugator activity for each trial. The change score was averaged for each picture category to provide a mean change in corrugator activity for each picture type.

Findings

The main result was that the magnitude of startle to acoustic probes delivered while the child was expecting a strong blast of air to the throat or while watching unpleasant pictures failed to differentiate the temperamental groups.

Startle to air puff. As anticipated, magnitude of startle and corrugator activity were significantly larger, by a factor of 3, to the light that warned of an air puff compared with the safe trials ($F\ 3/585 = 108.7$ for startle; $F\ 3/618 = 50.1$, $p < .001$ for corrugator activity; see Table 4.11). However, temperament made no contribution to the variation in magnitude of potentiated startle to the warning light (based on subtracting the values from safe trials from the values to the warning light) or to the baseline startles. Although startles while sitting in the dark were not different from magnitudes displayed during the baseline or safe periods, girls had larger startles than boys ($F\ 1/195 = 5.63$) but showed equivalent corrugator activity. Finally, low-reactives showed less corrugator activity as well as less potentiation of corrugator activity to the warning light than the other temperamental groups who displayed equivalent values ($t\ (210) = 2.33$, $p < .05$ for corrugator activity; $t\ (205) = 2.01$, $p < .05$ for potentiation of corrugator activity).

Although high-reactives did not have larger startles, we wondered whether children who showed very large startles early in the procedure might be different from those with smaller values because there was an obvious habituation of startle magnitudes dur-

Table 4.11. Mean startle and corrugator values for the puff procedure (microvolts).

Variable	Low-reactives		Others		High-reactives	
	Boys	Girls	Boys	Girls	Boys	Girls
Startle						
Baseline	17.4	19.8	13.1	22.3	15.5	17.7
Warning light	46.3	50.3	35.5	63.1	43.5	55.4
Safe	16.2	17.8	13.1	25.3	17.8	19.4
Dark	15.7	16.1	11.5	20.1	13.8	15.9
Corrugator						
Warning light	1.3	1.5	1.9	2.2	1.8	1.9
Safe	1.3	1.5	1.8	2.0	1.7	1.7
Baseline for warning	1.3	1.5	1.7	1.9	1.6	1.7
Baseline for safe	1.3	1.5	1.7	2.0	1.7	1.7

ing the last 8 of the 16 trials. We divided the distribution of mean startle to the first 4 warning trials into terciles (by sex) and examined the behavioral and physiological features of the children in each tercile.

To our surprise, the children who were spontaneous and relaxed with the examiner had the largest startles to the first 4 warning lights. Twice as many children with a rating of 1 (given to extremely uninhibited children), compared with a rating of 4, had values in the highest tercile for startle magnitude (a comparison of terciles 1 and 2 versus tercile 3 was significant, chi-square (1) = 3.9, p < .05). Hence, it is not surprising that children with the lowest fear scores in the second year also had greater startle potentiation. Thus, the most sociable, least fearful children had the largest startles to the first 4 warning lights. This result implies that these children might have had a higher level of cognitive engagement in the laboratory tasks, as suggested earlier. One index of involvement in the laboratory procedures is the recognition memory scores for the pictures seen during the ERP procedure. Children who were engaged should have better memory scores. We also suspected that extreme EEG asymmetry values, which imply a higher level of cortical arousal, might also accompany task engagement.

Fifty percent of the children who combined unusually high recognition memory scores (at least 46 of 48 correct) with extreme left frontal activation had larger potentiated startles than the 21 percent who combined recall scores lower than 46 with smaller left frontal activation values ($z < .10$). Greater involvement in the laboratory procedures was associated with larger startles to the threat of the air puff.

Higher beta power is also an index of cortical arousal. Because the correlation between startle magnitude and corrugator activity was low, it is possible that the children with high values on both might show greater cortical arousal. The distribution of beta power at rest was divided into terciles, and we examined the correlation between startle and corrugator activity to the warning light within each of the 3 levels of beta power. The correlation rose from a low value of .10 when beta power was low to a correlation of .47 when beta power was high. The more beta power at rest, the higher the association between startle and corrugator activity. However, the children with the greatest potentiation of both startle and corrugator activity were rated as uninhibited with the examiner, described by their mother as sociable, and reported that they liked new activities. The children with the greatest potentiated startle were sociable and exuberant, not anxious or defensive.

Startle to aversive pictures. The startle data to the 9 unpleasant pictures also failed to differentiate high- from low-reactives. Although startle and corrugator activity were larger to the unpleasant than to the pleasant pictures ($F\ 5/465 = 22.7$), and once again girls had larger startles than boys ($F\ 1/193 = 4.84$), there was no effect of temperament on startle or corrugator activity, and high-reactive boys failed to show potentiated startle to the unpleasant scenes (see Table 4.12).

Table 4.12. Mean startle and corrugator values for the pictures (microvolts).

Variable	Low-reactives		Others		High-reactives	
	Boys	Girls	Boys	Girls	Boys	Girls
Startle						
Baseline	16.1	21.2	13.1	29.5	13.9	16.7
Pleasant	13.1	13.2	11.3	15.8	11.3	14.7
Neutral	12.9	15.5	13.2	17.1	11.3	16.1
Unpleasant	14.6	14.8	14.1	25.6	10.8	16.2
Potentiation	1.2	1.1	3.7	9.6	−.40	.33
Corrugator (trial–baseline)						
Pleasant	.18	.56	.71	.39	.17	.20
Neutral	.26	.40	.69	.18	.29	.16
Unpleasant	.33	.84	.66	.89	.21	.37
Potentiation	.20	.48	−.04	.62	.00	.21

The average correlations for startle magnitudes or corrugator activity across the 3 types of pictures were high ($r = .60$ for startle; $r = .63$ for corrugator activity). However, as with the air puff, there was only a modest relation between startle magnitude and corrugator activity ($r = .20$). Moreover, corrugator activity better differentiated the unpleasant from the other pictures than did startle magnitudes. This result affirms the suggestion that activity in this muscle group is a sensitive index of the symbolic valence of events (Larsen, Norris, and Cacioppo, 2003).

Hence, we created 4 groups based on mean startle (across all 27 pictures) and mean corrugator activity to the unpleasant pictures (only unpleasant scenes elicited corrugator activity):

- Group 1: high startle/high corrugator
- Group 2: low startle/low corrugator
- Group 3: high startle/low corrugator
- Group 4: low startle/high corrugator

The Group 1 children showed the largest potentiated startles to the unpleasant scenes. The largest startles occurred when corrugator activity accompanied an eyeblink to an unpleasant picture.

Relations between startle to warning light and unpleasant pictures. The correlation across the puff and picture procedures was high for startle ($r = .56$) but low for corrugator activity ($r = .13$), because few children showed corrugator activity to the warning light. Hence, we created two groups based on magnitude of potentiated startle to the two aversive events. Group 1 was high on potentiated startle to both the aversive pictures and the warning light. Group 2 displayed low potentiated startle values to both incentives. More low- than high-reactives were in Group 1 (chi-square (1) = 5.4, $p < .01$); the finding for girls was in the same direction but missed significance (chi-square (1) = 3.5, $p < .10$). Further, the children in Group 1 were uninhibited with the examiner, while Group 2 children were the most inhibited. Only 4 of the 28 children who showed large potentiated startles to both incentives had been high-reactive infants. This result is exactly the opposite of what many psychologists would have predicted; it was the low-reactive, not the high-reactive, adolescents who showed the largest potentiated startles to the two aversive events.

Summary

Although startles to acoustic probes were larger when children anticipated or were watching an unpleasant event, there was absolutely no indication that high-reactive or inhibited children had larger potentiated startles than low-reactive or uninhibited children. Indeed, the evidence favors the opposite conclusion. This conclusion is supported by the fact that college students reporting a great deal of enjoyment from new activities, as do our low-reactives, showed their largest startles to the unpleasant pictures. Students who did not possess this trait did not show potentiated startles to the unpleasant scenes (Hawk and Kowmas, 2003). That is why we suggested that large startles to aversive incentives may not reflect a defensive or anxious state but a higher level of cognitive engagement with the procedure. Recall that it was the low-reactives who showed the orienting profile of a warm ear temperature but cool fingers. Preschool children who became highly engaged in a competitive game, who resemble our uninhibited children, showed a rise in cortisol to the competition, reflecting a deeper involvement (Donzella et al., 2000).

The complete corpus of evidence invites a critical perspective on the popular hypothesis that potentiated startle to events that are judged unpleasant reflects an aversive, anxious, or defensive state in the subject. It is more likely that startles are potentiated when an incentive provokes thought. The occasional appearance of a picture of a gun, snake, or bloodied soldier, intermixed with neutral and pleasant scenes, is unexpected and provokes an attempt to understand why the examiner is presenting these scenes. The individuals who brood the most show the largest startles. But a state of "cognitive effort" is different from, and relatively independent of, a state of fear, defensiveness, or anxiety.

Pictures of snakes with open jaws are not regular experiences, and neither children nor adults expect to see such scenes in a laboratory managed by a friendly adult who has thanked them for being so cooperative. Many psychologists regard a rise in heart rate to an event as an index of a fear state. But most individuals who show large startles to pictures of snakes and guns show a decrease in heart rate—which is a sign of surprise. A person who actually saw a snake on a forest trail would show a rise in heart rate because he was indeed fearful. People watch violent television shows because they are exciting or arousing, not because they create a fear of being harmed—most humans do not like to feel afraid of attack.

Biologists differentiate between in vivo and in vitro preparations because physiological activity is often different under the two conditions. Even atoms take on special properties under unusual conditions—for example, when temperatures approach absolute zero. Scientists who claim that pictures of dangerous animals or objects elicit a fear state have to prove that claim rather than simply assume its truth.

The concept *fear* has replaced the older, less evaluative but broader concept of *general arousal* to name the hypothetical state of a person who shows signs of cortical or autonomic activation to an incentive. One reason for this semantic substitution is the greater pressure on contemporary scientists to rationalize the relevance of their work for human pathology. Government agencies and private foundations are concerned, appropriately, with the prevalence of human anxiety disorders. Any scientific fact that might illuminate these symptoms is to be celebrated. It is not surprising, therefore, that investigators have interpreted potentiated startle as a way to measure variation in fear or anxiety. Unfortunately, the evidence fails to support this optimistic premise.

ANTHROPOMETRY

Rationale

We noted earlier a modest relation between body build, eye color, and temperament. A lean body build is more prevalent among patients with panic disorder (Bulbena et al., 1996) and among social phobics who profited least from social skills training (Kellett, Marzillier, and Lambert, 1981). Further, a small body size results in greater loss of heat because the surface-to-mass ratio is larger and, as a result, the noradrenergic system is more active in order to maintain thermal homeostasis.

Therefore, at the end of the battery of physiological tests, we measured each child's height and weight and measured the height (forehead to chin) and the width of the face at the bizygomatic, the broadest part of the face. The latter pair of measures was used to compute a ratio of width over height as an index of a broad or narrow face.

We also measured the child's eye color because of the modest association, in Caucasian children, between shy behavior and possession of blue eyes (Rosenberg and Kagan, 1987; Rubin and Both,

1989; Coplan, Coleman, and Rubin, 1998). Shy adults who are blue-eyed, compared with non-shy adults who have brown eyes, have a lower olfactory threshold for butanol. This fact implies a relation among amygdalar activity, olfactory thresholds, and eye color (Kagan, Herbener, and Little, 1987; Herbener, Kagan, and Cohen, 1989).

Of greater interest is the fact that blue-eyed, but not brown-eyed, adults show a significant rise in auditory threshold to tones following a 1-minute exposure to a 1,000 Hz tone at 120 db (Hood, Poole, and Freedman, 1976). This result suggests some compromise in the function of the basilar membrane or cochlear nucleus to increasing sound energy in blue-eyed individuals. This effect could be due, in part, to the fact that blue-eyed individuals have less melanin in the cochlea. There is a positive correlation between the concentration of melanin in the iris and the cochlear nucleus. Thus, eye color in Caucasians is associated with physiological features that might have psychological consequences.

Darker pigmentation of skin and iris occurs when the melanocytes produce eumelanin rather than phaeomelanin. An allele of the agouti gene is responsible for a protein that acts on receptors for melanocortin on the melanocytes to inhibit the production of eumelanin. This is one mechanism controlling the balance between the production of eumelanin and phaeomelanin. The agouti gene in rodents, which is very similar to the human version (Kwon et al., 1994), is associated with gray fur and more aggressive and/or defensive behavior. The non-agouti allele is associated with a dark black coat and tame behavior (Lu et al., 1994; Hayssen, 1997; Valverde et al., 1995). Hence, we hypothesized that Caucasian children with light blue eyes might be more inhibited than those with brown eyes (Xiao et al., 2003).

Findings
Of the current sample of 11-year-olds, 48 percent had blue eyes, 39 percent had brown eyes, and 13 percent had hazel or green

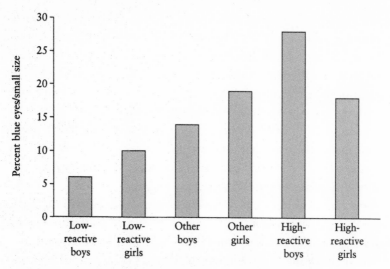

4.14 Percent of small blue-eyed children in each temperamental group.

eyes. Blue eyes were a bit more common among high- than low-reactives (54 versus 40 percent), but especially among high-reactive girls (55 percent of high-reactive girls versus 36 percent of low-reactive girls had blue eyes; chi-square (1) = 3.3, p < .10).

We combined height and weight, separately by sex, to create groups varying in body size. Children with height and weight values in the lowest tercile for their gender were classified as small; those with height and weight values in the highest tercile were classified as large. Three times as many high- as low-reactives combined a small body size with blue eyes (24 percent of high-reactives, 7 percent of low-reactives, 14 percent of others; chi-square (1) = 4.7, p < .05; see Figure 4.14). One of every 4 high-reactives, compared with 1 of every 14 low-reactives, and 1 of every 7 other children combined a small size with blue eyes. These results imply a genetic contribution to a cluster that combines size, eye color, and behavior (Song et al., 1994).

A relation between eye color and body size, on the one hand, and temperament, on the other, will not surprise those who study the changes in physical characteristics that accompany domestic breeding in mammals, especially wolves, foxes, mink, and cattle. Mice with the agouti gene are more fearful, more difficult to handle, have a different cranial and facial structure, and have a higher ratio of norepinephrine to dopamine in the brain stem than mice with the non-agouti allele. Adults with less fat and muscle than the average person, often ectomorphic in body build, are more accurate than others in detecting subtle changes in their heart rate (Cameron, 2001), and the primary sensory receptors within the cardiovascular system enabling this detection are located in the aortic arch. It may not be a coincidence that the aortic arch, the melanocytes that influence eye color, the sympathetic nervous system, and the maxillary bone of the face are all derivatives of the neural crest cells of the embryo (Hayssen, 1997). The neural crest cells are influenced by many molecules, including serotonin (Vitalis et al., 2003) and norepinephrine transporter (Ren et al., 2003). Thus, it is reasonable to speculate that the chemistry of the neural crest cells, before or after their migration, partly controlled by genes, contributes to the psychological differences between the two temperamental groups.

If minimal fear of novelty and extremely tame behavior are associated with distinct physical features in fox, rats, and mice, we should not be surprised that high- and low-reactive children differ in eye color and body size. It is possible that genes that mediate the time of migration of the neural crest cells, and perhaps the chemical properties of these cells, are pleiotropic and contribute to physical and behavioral features that differentiate between high- and low-reactives.

Homo sapiens sapiens is the most domesticated of the great apes: we are less fearful of strangers than are chimpanzees, gorillas, or orangutans. Comparisons of the domesticated and wild forms of many mammalian species—fox, pig, goat, and dog—reveal that the domesticated strain has, relative to the rest of the brain, a smaller limbic area. The central nucleus of the amygdala in particular is smaller in humans than in chimpanzees, and that may be one reason why we show less intense emotional reactions to strangers and other unfamiliar situations (Kruska, 1988).

Conclusion

With the exception of potentiated startle, the biological data were in modest accord with the expected outcomes for children who had been high- or low-reactive infants. The high-reactives were more likely to be inhibited in their behavior and show right hemisphere activation, a larger Wave 5, a larger Nc to the discrepant scenes, and greater sympathetic tone in the cardiovascular system. Chapter 5 attempts to integrate this behavioral and biological information.

5

INTEGRATING BEHAVIOR
AND BIOLOGY

Our seminal discovery was that a proportion of high- and low-reactive infants displayed theoretically expected behavioral and biological properties over a decade after their original temperamental classification. About 1 of every 4 high-reactives was fearful to unfamiliar events in the second year and, at age 11, subdued with the examiner, rated as shy by the mother, and reported disliking novelty and interaction with large numbers of children. About 1 of every 5 high-reactives, and 1 of every 3 low-reactives, preserved to age 11 both their expected behaviors and some appropriate biological features, but less than 5 percent of high- and low-reactives developed a combined behavior and biology characteristic of the complementary category (Figure 5.1).

Seven high-reactives (4 boys and 3 girls from a total of 68 high-reactives) were prototypic examples of their temperament. They were very fearful in the second year (one child was the most fearful 21-month-old in the sample), extremely quiet with the examiner, described by their mother as unusually shy, and displayed either right or extreme left frontal activation, a very large Wave 5, high sympathetic tone, and a large Nc to the discrepant scenes. No low-

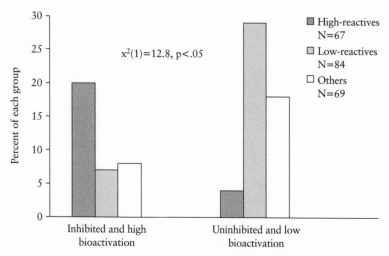

$x^2(1)=12.8, p<.05$

High-reactives
N=67

Low-reactives
N=84

Others
N=69

Percent of each group

Inhibited and high
bioactivation

Uninhibited and low
bioactivation

5.1 Percent of each temperamental group showing inhibited behavior
and high bioactivation or uninhibited behavior and low
bioactivation.

reactive child showed this pattern. These 7 high-reactives also displayed uncommon reactions during the 4-month battery. They
cried intensely to the olfactory stimulation and were much more
difficult to soothe than most infants who became temporarily upset. One boy showed long periods of motor spasticity and sucked
with extraordinary avidity when given a bottle during a short
break in the laboratory procedure.

The low-reactive boys were equally distinctive. Twelve low-
reactive boys—26 percent of all low-reactive boys and 10 percent
of all boys in the sample—earned the sobriquet "strong, silent, Clint
Eastwood types" because they were relaxed, quiet, never complained about the multiple electrode placements on their scalp and
face, and showed high vagal tone, low beta power, moderate left
frontal activation, a small Wave 5, and very low Nc values to the
discrepant pictures (see Figure 5.2). Their posture was defining, for
there was absolutely no muscle tension in their trunk or limbs

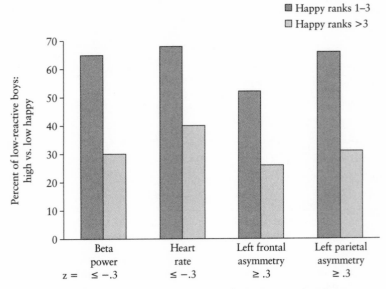

5.2　Percent of low-reactive boys with low beta power, a low heart rate, and left frontal and parietal activation as a function of describing themselves as "happy" or "less happy."

throughout the 3-hour battery. And they described themselves as "happy." No high-reactive displayed this profile.

It appears that the biology associated with maleness and the biology that enables low reactivity combined to produce a distinct pattern in about 1 of every 4 low-reactive boys. It is reassuring that 10 percent of the boys in a large sample of 4.5-year-old British children showed a similar profile. They were sociable with the examiner, unperturbed when their mothers left them alone in an unfamiliar laboratory room, and had very low heart rates—these are the strong, silent types (Stevenson-Hinde and Marshall, 1999).

However, to our surprise, a group of 7 low-reactive boys with very low values for beta power, Wave 5, and resting heart rate ($z <$ -0.5) behaved in different ways with the examiner. Three of them showed minimal emotion, low arousal, and infrequent conversation or smiling. But 4 of these boys talked and smiled frequently, often with energy and excitability. One boy, who talked continually in a loud voice, was exuberant. We believe that no one watching this boy would have guessed that his infant temperament and values on the biological variables at 11 years were similar to those of the 3 "strong, silent" boys.

As one might expect, combinations of biological variables separated high- and low-reactives better than any single measure. We used a standard score of ±.00 to designate high versus low values on each of the 7 variables that separated high- and low-reactives and computed a mean for combinations of these variables. This analysis was restricted to the children who had values on all 7 variables: right parietal activation in the EEG, Wave 5, temperature of the index finger, cardiac ratio, resting heart rate, magnitude of Nc to the first oddball stimulus, and the novel invalid pictures. The mean standard score across these 7 variables separated the high- and low-reactives better than any one measure (odds ratio = 7.0; chi square (1) = 13.1, $p < .01$). Seventy-eight percent of high-reactives, but only 34 percent of low-reactives, and 47 percent of others had a mean greater than .00 across the 7 variables. The low-reactive boys had the lowest mean, and the high-reactive boys the highest, across the 7 variables ($F\ 2/112 = 3.08$, $p < .05$ for the interaction of temperament by gender).

The children with a positive mean across all 7 variables—more often high-reactives—had their largest values on Wave 5. The children with a negative mean—more often low-reactives—had their largest values on the index of vagal tone (a high cardiac ratio combined with a low resting heart rate). This result supports our contention that the two groups do not differ quantitatively on a continuous physiological property; the biological foundations for high- and low-reactivity are qualitatively different. This suggestion is also supported by the fact that more high- than low-reactives combined a small body size, blue eyes, and infrequent smiling over the course of childhood.

It is interesting that low values on the biological measures, especially Wave 5 and the Nc voltages to the oddball and novel invalid pictures, were more characteristic of low-reactives than high values were for high-reactives. The ratio of low- to high-reactives (both sexes) with low values on these variables ($z < -.10$) was 10 to 1. It appears that low-reactives find it especially easy to attain a state of low biological arousal, while high biological arousal can reflect either a stable trait or a transient state produced by the laboratory interventions.

All animals must be capable of being aroused by threat or challenge. It is much easier to frighten children than to create a chronic state of relaxed spontaneity. Analogies exist in other domains. It is easier to heat objects to very high temperatures than to cool them to very low ones, and easier to increase the velocity of an atomic particle to very high speeds than to produce exceptionally low velocities. A Nobel Prize was awarded to a scientist who accomplished the latter feat.

Corroborating Data

These results are remarkably similar to those reported by Fox and colleagues (2001), who followed three different temperamental groups first classified at 4 months with a battery very similar to the one administered in this study. One group consisted of high-reactives; a second group resembled our low-reactives, for they were low in motor activity and crying and displayed very little vocalization and smiling. The third group was similar to our aroused category, combining high motor activity, frequent vocalization, and smiling with minimal crying. These three groups were observed at 14, 24, and 28 months, and EEG data were gathered at 9, 14, 24, and 48 months. Three findings are of special interest.

First, the high-reactives were most likely to be inhibited on all three behavioral assessments. The proportion of high-reactives with a consistent inhibited style (1 of 4) is very close to the proportion we found in our sample. Second, the consistently inhibited children showed right frontal activation at 9 and 14 months, and the high-reactives who were consistently inhibited were most likely to be right-frontal-active. Finally, the children who were consistently un-inhibited had shown high motor activity, vocalization, and smiling but minimal distress at 4 months. One-half of the children in this group were behaviorally uninhibited on every evaluation (Fox et al., 2001). The similarity between this evidence and our own is re-assuring.

Another longitudinal study, which began at 2.5 years rather than in infancy, is also corroborative. These middle-class Wisconsin children were classified as inhibited, uninhibited, or neither based on behavior with an unfamiliar child in a play situation. About one-third of these inhibited children remained inhibited when they were 7.5 years old. Most of the remaining inhibited toddlers moved to the middle of the distribution and were neither extremely sociable nor shy. About 40 percent of the original group of uninhibited children remained uninhibited; the rest also moved toward the middle; that is, very few became inhibited. These results, similar to our own, affirm that the probability of an inhibited child becoming uninhibited or an uninhibited child becoming inhibited is very low (Pfeifer et al., 2002).

About 10 percent of the infants in the Thomas and Chess longitudinal study ($N = 14$) were described by their parents as irregular in routine, irritable, and likely to avoid unfamiliar events. These infants, classified as difficult, shared some features with our high-reactive infants. The probability of developing symptoms during later childhood depended on the home environment (Thomas, Chess, and Birch, 1969). The difficult infants who were not over-protected by parents were least likely to develop symptoms—a result in accord with the research of Arcus (1991). The infants classified as easy, who resemble our low-reactives, were least likely to develop psychological problems, but the few whose parents permitted chronic disobedience were more likely to develop maladaptive behaviors.

Temperament Constrains

Despite the significant predictive relation between the infant classifications and the profiles at age 11, only about one-fourth of the infants in the high- and low-reactive groups actualized a behavioral and physiological profile in accord with theoretical expectation. However, very few showed the profile of the complementary group. Rather, most children displayed behavioral and biological patterns more characteristic of randomly selected middle-class Caucasian children, a result affirmed by the work of Fox, Rubin, and their colleagues. Apparently many high-reactives had learned to cope with their earlier tendency to become uncertain in response to unfamiliar events and had developed a persona that was not obviously shy or timid. Recall the essay written by the 13-year-old boy who had been a high-reactive infant and a very fearful 2-year-old which described how he conquered his timid persona through "will," even though he knew he did not yet possess the relaxed spontaneity of the typical low-reactive boy.

Thus, the most accurate summary of this evidence is that an early temperamental bias prevented the development of a contrasting profile. Put differently, the probability that a high-reactive infant would not become an ebullient, sociable, fearless child with high vagal tone, small Wave 5, a small Nc, and left frontal activation is very high. But the probability that this class of infant would become extremely shy and show high sympathetic tone, a large Wave 5, right parietal activation, and a large Nc is much lower. And the prediction that a low-reactive infant will not become a very shy 11-year-old with high biological arousal is much more certain than the prediction that this child will be exuberant and show low biological arousal. An infant's temperament was more effective in constraining the development of certain profile than in determining a particular profile.

The principle that a temperamental bias eliminates many more possibilities than it determines applies to environmental conditions as well. If all one knows about a sample of children is that they were born to economically secure, well-educated, nurturing parents, it is much easier to predict what they will not become—criminals, school dropouts, psychotics, drug addicts, homeless, or impoverished—than to predict what they will become. Predictions of the specific characteristics that will be part of the adult's personality are unlikely to be correct.

Similarly, among children born in poverty to single parents who did not graduate high school, it is much easier to predict the occupations these children are likely *not* to have when they grow up—museum curator, archeologist, Wall Street stock broker, cellist—than those they will. The final fate of a neural crest cell in a 6-week-old embryo, whether sensory ganglion, melanocyte, or heart muscle, is less certain than the fact that this cell will not become connective tissue, gut, or part of the reproductive system.

Thus, a temperamental bias can be likened to the basic form of the song of a particular species of bird. The animal's genome constrains the basic form of the song but does not determine all of its variations because the adult song depends on exposure to songs of conspecifics and the opportunity to hear its own vocal sounds. A bird deafened early develops an abnormal song. Thus, knowing that a bird is a finch rather than a meadowlark allows one to predict with great confidence the songs it will *not* sing but permits a far less certain prediction of the particular songs it will sing (Brainard and Doupe, 2002).

Consider one more illustration of the constraining power of initial conditions: Imagine a stone rolling down a steep mountain over a 5-minute interval. An observer can eliminate a great many final locations after each 10 seconds of descent, but it is not until the final second that she will be able to predict exactly where the stone will come to rest. All information eliminates alternatives. A listener who hears, "The boy ate . . ." knows many, many words that will not occur next, while remaining uncertain as to which particular word is about to be spoken. When the high promises of the genome project are met, and parents can request a complete genomic analysis of their newborn, an expert will be better able to tell parents what the infant will not become—schizophrenic, bipolar, talented athlete, creative composer, or brilliant mathematician—than to inform them about the characteristics their infant will possess two decades later. The discovery that high-reactive and low-reactive temperamental biases seriously constrain the envelope of possible outcomes, while not determining a particular cluster of features, may be our most significant finding.

Who Is Being Compared?

Most of our analyses compared high- and low-reactive children, rather than comparing high- or low-reactives with children in another group. For example, high-reactive girls showed greater right parietal activation than low-reactive girls, but the low-reactive girls had asymmetry values similar to those of the other girls. A meteorologist, when asked about the causes of tornadoes, assumes a contrast between tornadoes and everyday weather, rather than a comparison of tornadoes with hurricanes. The reasons for a tornado rather than a hurricane are different from the reasons for a tornado rather than a warm cloudless July day. The reasons why 1 percent of American families are very wealthy and 20 percent are poor pertain to conditions that are not relevant when we try to explain why 1 percent are wealthy and over 50 percent earn between $25,000 and $75,000 annually. The answer to the first comparison—between the very wealthy and the poor—requires reference to the family's social class, years of education, and ethnicity. These three factors are less important in the second comparison—between the very wealthy and those with average income—because most Americans who earn between $25,000 and $75,000 had middle-class parents, attended college, and are Caucasian, just like most wealthy Americans. To understand why a small number of

Table 5.1. Profiles of biological variables for low-reactives, high-reactives, and others for standard scores of ± 0.5
($-$ = < -0.5; + = > 0.5).

Variable	Low-reactives	High-reactives	Others
Wave 5	$-$	$+$	$-$
Heart rate	$-$		$+$
Nc oddball	$-$	$+$	$-$
Beta power	$-$	$+$	$+$
Warm fingers	$-$	$+$	$+$

families are wealthy and a majority have moderate incomes, we have to take temperament, parental socialization, and luck into account.

High- and low-reactives are only two of the large number of temperamental categories scientists will explore in the future. The group we called *other* (composed of the distressed and the aroused groups) displayed biological and behavioral profiles somewhat different from the high- and low-reactives (Table 5.1). The 4-month-old boys who showed low motor activity but high distress were shy and subdued at age 11, but they did not show a large Wave 5 or large Nc voltages to the invalid scenes. The girls in this group were also shy at age 11 but did not show the high baseline heart rate or right parietal activation characteristic of high-reactive girls. The 4-month-old boys with high motor activity and minimal crying (the aroused group) had a small Wave 5, resembling low-reactive boys, but a higher heart rate and extreme, rather than modest, left parietal activation at age 11. The girls in this group differed from low-reactive girls in showing high beta power and very warm finger temperatures.

The Modulating Effect of Gender

Although the mean across all 7 biological measures differentiated the two reactivity groups for both sexes, the variables that best

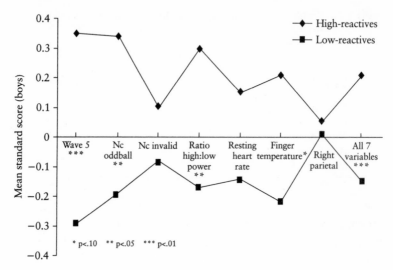

5.3 Mean standard score for major biological variables for high- versus low-reactive boys.

separated high- from low-reactives differed for boys and girls (Figures 5.3 and 5.4). Wave 5 best separated high- from low-reactive boys, followed by the magnitude of Nc to the oddball flower, cardiac ratio, and finger temperature. The high- and low-reactive girls were separated most clearly by right parietal activation, Wave 5, and the Nc response to the novel invalid scenes. Right parietal activation was the most distinctive characteristic of high-reactive girls, and the high-reactive girls with this feature were most likely to have displayed anxious symptoms when they were 7 years old.

The greater prevalence and more robust preservation of inhibited behavior in high-reactive girls than high-reactive boys is in accord with evidence from an independent study of a normative sample assessed at 14, 20, and 32 months, for the consistently inhibited children were more often girls than boys (Reznick et al., 1986). Women report a greater fear of the bodily sensations that accompany anxiety than men (Zvolensky et al., 2001), and an MRI analysis of the brains of healthy adults revealed that the ra-

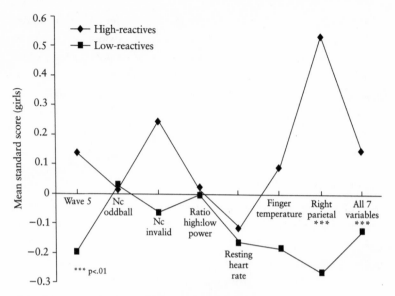

5.4　Mean standard score for biological variables for high- versus low-reactive girls.

tio of the volume of the orbitofrontal prefrontal cortex to the amygdala was higher in women than in men (Gur et al., 2002). Castrated male rats with low androgen levels showed a larger increase in ACTH, cortisol, and c-fos activity in area CA1 of the hippocampus to a novel experience than normal males. This result suggests that male sex hormone mutes the reactivity of the HPA axis and hippocampal neurons to unfamiliar events (Kerr, Beck, and Handa, 1996).

The higher level of sympathetic arousal in girls, compared with boys, may contribute to the frequently reported finding that affiliative relations appear to be more central to women than to men (Stroud, Salovey, and Epel, 2002; Stroud et al., 2000). Rats or monkeys who encounter an unfamiliar event, or are in an unfamiliar place, show lower sympathetic arousal and fewer behavioral signs of fear if a familiar conspecific is present than if they are alone in the unfamiliar environment. If more females than males are at a high level of sympathetic arousal, friendships may serve to mute the arousal. Low-reactive boys, who often possess high vagal tone, may seek unfamiliar experiences because such events do not provoke high sympathetic arousal. It may be relevant that male mice (from the C57 BL/6 strain) were more likely than females to explore a novel object encountered in an area familiar to them (Frick and Gresack, 2003).

The Role of the Amygdala

We believe, but have not yet proven conclusively, that the physiological bases for the properties of high- and low-reactives rest, in part, on the neurobiology of the amygdala, and perhaps the bed nucleus, and their projections to cortex, striatum, hypothalamus, brain stem, and the autonomic nervous system. The biological measures that best differentiated high- and low-reactives are indirect signs of amygdalar excitability.

Amygdalar priming of the inferior colliculus should potentiate Wave 5 values. Amygdalar projections to the sympathetic nervous system should be accompanied by a lower ratio in the cardiac spectrum, a higher heart rate, larger heart rate acceleration to challenge, and warmer fingertip temperatures. Amygdalar projections to the basal nucleus of Meynert and the brain stem nuclei that excite cortical neurons should potentiate Nc voltages to the discrepant pictures. Finally, asymmetry of amygdalar projections to the cortex could create more extreme asymmetries of cortical activation (Halgren, 1992; La Bar et al., 1998).

It is easy to find an alternative explanation for each biological difference between high- and low-reactives that does not rely on amygdalar excitability. But the fact that several biological measures separated the high- and low-reactives makes our interpretation a bit more compelling. The diagnosis of an illness provides an analogy. Although high fever, chills, and a high white cell count could be due to a variety of causes, this combination of physical symptoms in a person who has just returned from a rural seaside village in Honduras makes it highly probable that malaria is the culprit. Similarly, the soundness of our theoretical argument rests on the entire corpus of evidence, not on the strength of any one finding.

If high values on several of the biological variables imply a more excitable amygdala, children with high values should dislike excessive novelty in their daily lives, while those with low values should be attracted to new experiences. Four items on the child Q-sort reflected the attitude toward novelty. The items were: "I like to be first in gym class to try playing a new game"; "I like to play with many children at recess"; "I like going on roller coasters at an amusement park"; and "I like going to new places." We correlated the mean rank across these 4 items with the mean standard score for the 7 biological variables; a positive correlation meant that high biological arousal was related to dislike of new experiences. The correlations were positive for high- and low-reactive boys and low-reactive girls (average $r = 0.3$, $p < .05$) but not for high-reactive girls ($r = .00$), despite no significant differences between these temperamental groups in the mean rank across the 4 Q-sort items.

More important, 47 percent of low-reactives, but only 15 percent of high-reactives, combined a preference for novelty with low biological arousal; 42 percent of high-reactives but only 23 percent of low-reactives combined a dislike of novelty with high biological arousal (based on division at the mean, chi-square (1) = 17.0, odds ratio = 5.6). One of every 2 low-reactives, but only 1 of 7 high-reactives, said she liked new experiences and, in addition, showed low biological arousal. When low-reactive children reported a preference for novelty, the probability was 0.8 that they would show low biological arousal. When high-reactive children disliked novelty, the probability was 0.8 that they would show high biological arousal. These results support the suggestion that one consequence of amygdalar excitability is a reluctance to seek unfamiliar experiences regularly. It is relevant that adults categorized as inhibited in the second year showed greater amygdalar activity to unfamiliar faces than those classified as uninhibited (Schwartz et al., 2003a). However, it is conceivable that maintaining a bold or cautious persona over the childhood years could change the threshold of responsivity to the neurotransmitters or modulators that influence the amygdala. This phenomenon occurs in animals. The responsivity of the neurons controlling the tail-flip in crayfish following an increase in available serotonin depends on whether the animal had recently won or lost a fight with another crayfish (Yeh, Fricke, and Edwards, 1996).

Relation of Biology to Behavior

The relation between each of the biological variables and the child's social behavior at age 11 was low, except for the cardiovascular variables. A longitudinal study of British children also reported that the children who preserved their inhibited behavioral style from 4.5 to 7.5 years of age had higher heart rates than others (Marshall and Stevenson-Hinde, 1998).

Further, the occurrence of a significant relation between biology and behavior was often restricted to a particular temperament or gender group. For example, the correlation between the mean of the 7 biological variables and the mother's rating of "shyness" was positive for low-reactive girls ($r = .45$, $p <. 05$) but negative for high-reactive girls ($r = -.45$, $p <. 05$). Low-reactive girls with high biological arousal were perceived by their mother as relatively shy, compared with other low-reactive girls. But high-reactive girls with equally high levels of biological arousal were described as less shy than the other high-reactive girls.

The Wave 5 data provided a second example. We compared children with high or low Wave 5 values ($z = \pm.10$) with respect to other biological measurements. Low-reactive boys with a large Wave 5, compared with low-reactive boys with a small Wave 5, displayed high beta power, a warmer index finger, and greater sympathetic tone. But these differences did not occur for any other temperamental group. Similarly, high-reactive girls with a large Wave 5, compared to those with a small Wave 5, showed greater right parietal asymmetry, but this difference, too, was absent for the other groups. Boys with left hemisphere activation at both frontal and parietal sites had low beta power; girls with the same EEG profile did not vary in beta power. Thus, low correlations among the biological measures across all the children are due, in part, to the fact that the pattern of relations varies by temperament and/or gender. That is why David Magnusson (2000) has argued that a single variable should not be studied in isolation.

A small number of low-reactives were as emotionally subdued with the examiner as the larger group of high-reactives. Observers watching the films of these quiet, serious children would have difficulty deciding which had been high-reactive and which low-reactive as infants. But the subdued high-reactives, compared with the equally subdued low-reactives, showed more extreme asymmetry of frontal activation, right parietal activation, a larger Nc to the oddball picture, and a larger Wave 5.

High- and low-reactives who were described by their mothers as highly sociable also differed in their biological profiles. The 36 sociable low-reactives showed a smaller Wave 5, greater left parietal activation, and greater vagal tone than the 24 high-reactives rated as equally sociable. When clinicians add biological information to the interview data they now rely on to arrive at a diagnosis, they will discover that most current mental illness categories are heterogeneous, and patients with the same diagnosis vary in their physiological functioning.

The nonlinear relations between measures are a second reason for the complexity of biological and psychological relations. For example, the relation between the mean standard score for the 7 biological variables and the rating of degree of inhibition with the examiner was nonlinear. The children with a rating of 1, which reflects a maximally relaxed, spontaneous child, had low values, but the remaining children with rating 2, 3, or 4 had similar values on the composite index of biological arousal. Ten of the 17 extremely uninhibited children with a rating of 1 and a mean biological index score of less than −0.3 had been low-reactive; only 1 of the 17 had been a high-reactive infant.

Many social scientists assume that, given a well-defined context, a particular value on a variable has the same meaning across subjects and should have the same pattern of correlates. Many statistical procedures require that assumption. However, a high body temperature can be associated with a high heart rate and fatigue in a person who has exercised for several hours on a very warm day or with a high white cell count in patients with a bacterial infection. Most measurements are ambiguous until the investigator specifies relevant characteristics of the agent; temperament and gender are two such characteristics.

The relative independence of the biological and behavioral variables raises an important theoretical issue. What construct should we use to categorize a child who showed high values on the biological variables but was not unusually shy, timid, or subdued? More high- than low-reactives had greater activation in the right parietal area, larger Wave 5 values, and larger increases in beta power in response to challenge. However, many high-reactives with this profile were not very different behaviorally from high-reactives who did not show this biological pattern.

These data imply that some brain states may have minimal influence on the quality of a person's consciousness, intentions, or behavior, just as some alleles have no implications for either phenotypic features or adaptation (Tang et al., 2001). Panic patients inhaling carbon dioxide (35 percent concentration) showed the expected rise in heart rate and blood pressure and reported feeling distress. But control subjects breathing the same concentration of carbon dioxide, and displaying the same cardiovascular profile, failed to report any distress—their biology and their subjective state were dissociated (Schmid et al., 2002). A biological state represents only a potentiality for a psychological property.

There is an increase in order as we move from neuronal excitation in the amygdala and ventral striatum to the limb movements of the 4-month-old infants to the mobiles and voices because there are fewer possible ways an infant can move its arms and legs than possible states of the underlying neuronal ensembles. There is some loss of determinism at the junction between the neural and behavioral domains. It is not possible, for example, to predict the exact vigor and duration of a child's limb movement from complete knowledge of the immediately preceding neuronal profile.

Consider a rain shower in San Francisco. Where shall we begin our description of this event? Should we start with the envelope of moist air over the Pacific on the previous day, the shape and moisture content of the cloud formations over San Francisco, or the first raindrops? The components of each phenomenon require a special vocabulary. That is, the words describing the moving air mass over the Pacific (size, moisture content, velocity, and temperature) are inappropriate for the shower (duration and density of raindrops per hour). At each boundary or transition point in a dynamic series, the degree of determinism is compromised as some uncertainty is introduced. Two air masses of identical size, moisture content, and temperature 100 miles west of San Francisco on two different days will encounter nonidentical conditions and lose their similarity minutes after striking the California coastline. As a result, each will produce different showers.

Although the statistical independence of most biological measurements is a common result across many laboratories, this fact remains puzzling. One reason is that each target, whether heart, inferior colliculus, capillary, eye muscle, or neuronal ensembles in frontal lobe, has local control mechanisms that are independent of the central processes an investigator wishes to quantify. Although the amygdala affects the heart, circulatory vessels, and inferior colliculus, directly or indirectly, each site is also modulated by local processes within the structure that can mute or enhance the effect of the amygdalar projections. The amount of commercial activity in 30 banks across America provides an analogy. Although the decisions of the Federal Reserve Board in Washington influence bank activity in all 50 states, there are probably low correlations, on any particular day, among the levels of commercial activity in Chicago, Boise, Topeka, Los Angeles, Skowhegan, and Trenton because of local economic conditions in each city.

Disease categories provide a more relevant analogy. Consider a large sample of adults who have high body temperature, weight loss over the past 12 months, and self-reported fatigue at the end of each day. The correlations among these three dimensions in a large, unselected sample is close to zero because each of the three features has several independent causes. However, because an infection with tubercle bacilli is associated with all three symptoms, there will be a small group of individuals with all three features who belong to a special category. Stated formally, if each of a set of measures has more than one origin, and the origins are independent, the correlations among the measures will be low even though one origin is common to all of the measures for a small proportion of a sample.

The independence of measurements presumed to index the same construct poses a problem. Research would be easier if scientists could rely on one or two variables to index a complex psychological construct like an inhibited temperament. Unfortunately, the multiple controls on each measurement imply that a high or low value rarely has a univocal meaning. Recall that both boisterous and timid children display high cortisol levels, but in different contexts and at different times of the year (Gunnar, 1994).

A final reason for the independence derives from the fact that the child's thoughts and feelings at the time of measurement, which can be unrelated to the investigator's intention, can influence brain activity and therefore reactivity of varied targets. Most of the time the pattern of brain activation, as measured by fMRI or PET, varies with the cognitive operations required by a task, rather than with the stimulus in the perceptual field or the required motor response. That is, biological measurements in an awake human are often influenced by the person's thoughts during the procedure.

For example, adults differ in preferred style of decision-making on cognitive tasks; some are cautious, others are impulsive. These two personality types produce different ERP waveforms when presented with words that might, or might not, have been seen on a prior list (Windmann, Sakhaut, and Kutas, 2002). Perhaps some low-reactives showed right hemisphere activation because they were imagining what their brain waves looked like while we were gathering the EEG data. By contrast, some high-reactives might have shown left hemisphere activation because they were thinking about why they had agreed to cooperate.

Behavioral variables presumed to index the same construct are also often independent. Rats or mice who show immobility in response to a conditioned stimulus that signifies an aversive event do not necessarily show potentiated startle to the same stimulus (Richardson and McNally, 2003). A child who is shy with unfamiliar peers might not show equally timid behavior in the face of physical challenges. As with the biological measures, each category of behavior is modulated to some degree by factors unique to its structure.

The history of psychology is littered with examples of reliance on a single dependent variable to index an abstract construct. Over a century ago, Francis Galton used a fast reaction time to measure human intelligence. A generation of behaviorists regarded the number of trials a rat required to learn a maze as an index of learning ability. Still others used the galvanic skin reflex or a rise in heart rate as a measure of arousal. Behavior in the Strange Situation is regarded by some psychologists as a measure of an infant's security of attachment. A rat's reluctance to explore the brightly lit alley of an elevated T maze is treated as an index of fear. And an increase in attention of 1 or 2 seconds is regarded as a measure of the infant's ability to add numbers.

In each of these examples, investigators did not first explore the multiple factors that could affect the measure they had chosen. Had they done so, they would have learned that each had more than one determinant, and therefore the inferences drawn from their evidence were vulnerable to critique. For example, many behavioral biologists treat the ultrasonic vocalizations of rat pups taken from their nest as reflecting a state of "emotional distress." However, the cooler temperature of the separated infant, caused by removal from the warm nest, leads to greater pressure within the abdomen, increased venous return to the heart, and ultrasonic vocalizations (Blumberg and Sokoloff, 2001). Attributing an emotional state to the rat pup does not seem to have any theoretical advantage.

It is rare for any current measure in psychology or biology to have a single cause for its distribution of values. Therefore, investigators who rely on one procedure to index a construct should first probe the varied conditions that affect that measure and then either control for unwanted causes or, if that is not possible, combine the measure with others so that a profile of scores becomes the dependent variable. To illustrate, when adults were exposed to a sudden loud sound (to produce an eyeblink), the magnitude of their startle was equally large when they were watching unpleasant pictures or solving difficult cognitive problems (arithmetic or anagrams). However, when large startles were combined with a measure of heart rate deceleration and corrugator activity, the unpleasant pictures provoked a very different profile than the cognitive tasks (Sorensen, McManis, and Kagan, unpublished).

The advantages of quantifying several variables simultaneously is seen in a study in which investigators performed a microdialysis of changes in dopamine in 3 sites—prefrontal cortex and the shell and core of the nucleus accumbens—in response to 5 different orally administered stimuli with different taste qualities. Although all 5 tastes produced increased dopamine in the prefrontal cortex, only the sweet taste that was unfamiliar (a combination of sucrose and chocolate) elicited increases in dopamine in all three sites. The aversive tastes (quinine and salt) produced dopamine activity in the prefrontal cortex and the core of the accumbens but not in the shell. Because these investigators measured changes in dopamine in several sites, they were able to differentiate among the different tastes. Had they only measured dopamine change in the prefrontal cortex, they would have concluded that all the tastes had the same effect (Bassareo, DeLuca, and Di Chiara, 2002).

If investigators can eliminate all the determinants except the one of interest, which is easier in physics than in psychology, they can come closer to evaluating the validity of a hypothesis. However, we could not stop our children from generating idiosyncratic thoughts during the laboratory session, and this fact could have contributed to the statistical independence of the biological variables. The potential influence of the child's thoughts can also explain the potentiated startle data.

Potentiated Startle

The most important null result was the absence of any relation between magnitude of potentiated startle to the two aversive events and either infant temperament or the behavioral signs of anxiety, fear, shyness, or timidity. Indeed, low-reactive children had slightly larger potentiated startles than high-reactives, a result consonant with the fact that adult extraverts showed larger startles than introverts to emotionally arousing pictures (Petren et al., 2002).

We suggested in Chapter 4 that magnitude of startle is often a function of the richness of ongoing thought when the acoustic probe occurs. Subjects have larger startles when they try to enhance their emotional reactions to unpleasant pictures (Jackson et al., 2000), imagine deeply personal memories with a positive valence (Miller, Patrick, and Levenson, 2002), anticipate a monetary reward (Skolnick and Davidson, 2002), or talk about emotionally arousing situations (Zech, Bradley, and Lang, 2002). Magnitude of startle is stable over two sessions when subjects see different pictures on each session but not very stable when they see the same pictures both times because less thought is provoked on the second session (Larson et al., 2000).

The low-reactive children may have been more engaged by the laboratory procedures than the tense high-reactives. Recall that more of the former showed orienting reactions. We suggest that a picture of a snake and the appearance of the light that warned of an air puff to the throat generated more thought among the relaxed, engaged children, and therefore larger startles. We suspect that many high-reactives, especially those who had been highly fearful in the second year, resisted complete psychological investment in all the laboratory procedures and, as a result, showed smaller startles.

The failure of potentiated startle to separate high- from low-reactives is reminiscent of our failure 3 years earlier to find any temperamental contribution to differences in magnitude of Stroop interference to threatening pictures. As with the potentiated startle data, the low-reactive uninhibited children were a bit more likely to show interference than the high-reactive inhibited ones. We had also gathered Stroop interference data on an independent group of 193 children, ages 4 to 8 years, who were subjects in a cross-sectional study evaluating the differences among 4 groups of children: those with a parent who had panic disorder, depressive disorder, panic co-morbid with depression, or no psychiatric symptoms. The children saw outline drawings of familiar objects—the outline for each picture was in 1 of 3 colors for younger children and 1 of 4 colors for older children. The pictures were symbolic of threat, a joyful mood, or were neutral in affective connotation. Each child had to name the color of the outline and ignore the picture's content.

Although the pictures signifying threat produced the most interference across all children (as expected), the variation in interference was not related to parental pathology. The pictures that produced the greatest interference were familiar objects with simple semantic names (for example, knife, snake, gun), suggesting that these pictures provoked a richer semantic network. But the extensiveness of the semantic network activated by a picture was unrelated to the psychiatric status of the children's parents.

Scientists have two choices when faced with our evidence on potentiated startle. The theoretically conservative position holds that potentiated startle does measure vulnerability to a state of anxiety, defensiveness, or uncertainty, and high-reactives are simply not more defensive. That claim is difficult to defend, but it is not bizarre. On the other hand, these data should motivate psychologists to question the popular interpretation of potentiated startle. The suggestion that cognitive processes may be more relevant to potentiated startle than affective states is rendered more credible if readers imagine how they would feel if, while sitting in a comfortable chair watching television, a cobra and gun barrel appeared in succession on the screen. We suspect that most would experience a rush of associations, but few would feel anxious.

Extreme Scores

There were many occasions when values in the top and bottom quartiles or terciles of a distribution differentiated the temperamental groups but mean values did not. For example, high-reactives showed extreme left frontal activation, while low-reactives showed modest left frontal activation. Sixty-eight percent of high-reactives, but only 35 percent of low-reactives, had standard scores of ±0.5 (or larger) for 3 or more of the 7 differentiating biological variables. Every high-reactive had an extreme value ($z = \pm 0.5$) on at least one of the biological measures, compared with only 50 percent of low-reactives (chi-square (1) = 11.6, $p < .01$).

The small group of children with extreme scores on several biological variables had an early childhood profile that was different from most other children. Seven high-reactives, but not one low-reactive, had high standard scores on 4 variables ($z > 0.5$): right parietal activation, Wave 5, Nc to the novel invalid scenes, and finger temperature. Every one of these 7 was extremely difficult to soothe during the 4-month assessment and was subdued at 14 and 21 months. A few of these children showed unique profiles on the early assessments. One 4-month-old girl continually turned her head away from the mobiles as if she were trying to avoid seeing the stimulus. One boy who clung to his mother throughout the 14-month-old laboratory evaluation brought a stuffed toy to the laboratory at 11 years of age—no other 11-year-old brought a toy for security. Thus, selection of children with extreme scores revealed intriguing relations that were undetected when we examined mean values across all high-reactives.

The heritability of behavioral inhibition in a large longitudinal sample of identical and fraternal twins in the second year revealed much higher heritabilities when the analyses were restricted to children with scores greater than 1 standard deviation from the mean (Manke, Saudino, and Grant, 2001). Adults with extreme values on self-reported shyness (the top 10 percent of a sample of over 2,000 individuals) were at greater risk for social phobia, but not those whose shyness values were in percentiles 40 to 60 (Chavira, Stein, and Malcarne, 2002).

In addition, on many occasions the correlations between or among variables were low across the whole sample, but the children with values at either extreme were different (see Figure 5.5 for an example). For example, despite no significant relation between the magnitude of Wave 5 and magnitude of startle to the warning light across all children, the 21 children with very small startles (the bottom 13 percent) had significantly lower Wave 5 values than the 14 children with very large startles.

Only 4 percent of a large sample of boys from different laboratories were persistently aggressive throughout the childhood years (Brody et al., 2003). Hence, the correlations between early and later asocial behavior across the whole sample would have been close to zero. Similarly the small positive correlation between a child's age (range of 6–18 years) and the adjusted area of the corpus callosum was due entirely to 7 percent of the sample who were

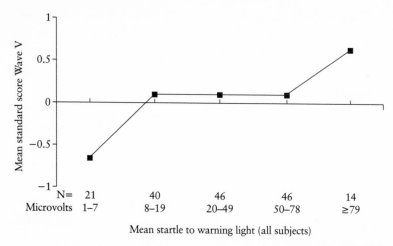

5.5   Relation between Wave 5 and startle to the warning light.

very young and had small callosal areas and 4 percent of the sample who were the oldest and had large areas. When these subjects were eliminated, there was no relation between age and the area of the corpus callosum for about 90 percent of the sample (De Bellis et al., 2001). Male vervet monkeys who were at either extreme on behavioral measures of impulsivity had a lower social rank than the large number of animals whose scores were in the middle of the distribution (Fairbanks, 2001).

This issue is of special relevance to social scientists who place great faith in average values. Early nineteenth-century scholars believed that the mean reflected the essence of a phenomenon and that any other value was less than perfect. Contemporary statistical practices accept this premise implicitly, and most reviewers of technical papers suspect any conclusion that is not based on mean scores. This bias is irrational because nonlinear functions are common in psychology, and current statistical procedures, especially analysis of covariance, often fail to reveal important relations. These procedures assume that the processes mediating a distribution of scores are the same for all values and the causal mechanisms vary only in magnitude. A reluctance to acknowledge the utility of examining extreme groups that might be qualitatively different from the rest of the sample is slowing progress.

Extreme values are often overdetermined by the unique and stable properties of the subject. Therefore, all are not equally likely to display extreme scores. For example, in 2002 and 2004 the New England Patriots won the Superbowl in the closing seconds of the game in a similar way—the same quarterback threw several successful passes that placed the team close enough to the rivals' goal to permit the same place kicker to make a winning field goal. This pair of highly improbable events occurred because of the special properties of this particular team and was not equally probable for all teams in the National Football League. Analogously, the biology of some high- and low-reactive adolescents made it likely they would display extreme values on some of our biological measures, and those who did had unique qualities 10 years earlier.

When the relation between two measures is nonlinear—a common occurrence in the life sciences—new qualities emerge at a transition value that can represent a distinct phenomenon. The behavior of a single ant (or a few ants) appears to be random. But when the number of ants in a colony reaches a critical density, coherent rhythmic activity emerges. A large colony of ants has unique qualities that cannot be predicted from, or explained by, an additive model that sums the activity of increasing numbers of ants and computes a mean amount of rhythmic activity across ant colonies of varied sizes.

In the final chapter, we will entertain some more speculative implications of this corpus of information.

6

Every new fact wrested from nature's closely guarded box of treasures elicits one reaction from the investigators, but quite a different one from the larger community and the institutions that supported the research. The fortunate scientists feel satisfied because their labor of love has placed a few new bricks on the jerry-built edifice called knowledge. But the larger community, wishing for more, expects the investigators to unbutton a little and to speculate on the broader significance of their constrained, statistically defended claims. This final chapter tries to respond to this reasonable request.

Temperament and Feeling Tone

The infant temperaments we call high- and low-reactive cast long shadows that changed their shapes over the course of development. The public sociability and obvious indifference to unfamiliarity displayed by low-reactives during early childhood typically evolved into more subtle features, especially a sanguine mood and a relaxed muscle tone. These adolescents experienced delight from new sights, sounds, and conversations that tweaked their understanding of the world and challenged habits that had become familiar. Most high-reactives, by contrast, replaced an obvious caution to unfamiliarity present during early childhood with private feelings of tension when anticipating social encounters and concern over the day's responsibilities as they approached the adolescent years. Many of these youth wished they were more relaxed and free from worrying about the critical thoughts others might entertain about them.

Two adolescent girls who had been high-reactive infants and inhibited in the second year were interviewed at home when they were 15 years old. Although both were relaxed and minimally defensive, they said they felt uncomfortable in crowds, concerned with the opinions of their peers, and exceptionally uneasy when they violated one of their personal standards.

Carl Jung's descriptions of the introvert and extrovert, written over 75 years ago, apply with uncanny accuracy to a proportion of our high- and low-reactive adolescents (Jung, 1961). We believe that each adolescent's dominant feeling tone, not the degree of shyness or sociability in his outward persona, is the seminal property that differentiates these temperamental groups in adolescence, as it did the poets T. S. Eliot and Sylvia Plath, on the one hand, from e. e. cummings and Robert Graves, on the other. The dysphoric quality in the poetry of the former pair is easily distinguished from the celebration of life and love in the poems of the latter two writers.

A personality type is a pattern of traits, each determined by a combination of temperament, personal experience, and the contexts of daily life. Temperament makes a more substantial contribution to feeling tone than to the public personality during adolescence and adulthood. The development of an individual, therefore, can be likened to a road map outlining different routes to each of many locations, with some itineraries more easily traversed than others. The developmental journey that leads to a relaxed or a tense feeling tone requires a more substantial contribution from temperament than does a sociable or shy posture with others.

The quality of a person's mood, whether chronic or provoked by an interruption, is influenced by a circuit that connects heart, blood vessels, muscles, and gut to the medulla, amygdala, and orbitofrontal prefrontal cortex (Damasio, 1994). Activity in peripheral body sites is transferred to the frontal cortex to produce, on some occasions, a conscious perception of a change in feeling tone. When that change is subtle, mildly unpleasant, and ambiguous in origin, the person might articulate his feeling as shame, guilt, regret, illness, fatigue, or, perhaps, possession by the devil. Some individuals experience these dysphoric changes more regularly or more intensely than others, and we believe that temperament makes a contribution to this subjective state. High-reactive adolescents, who we claim have a more excitable amygdala, live with a more reactive sympathetic nervous system, a higher level of muscle tension, and a greater cortical arousal to new experiences. Hence, they should be susceptible to more frequent, or more salient, evocations of an emotion that, in our culture, invites an interpretation of a personal flaw.

"Happiness" is the most frequent answer American youth give to the question "What do you want in life?" But happiness is elusive for those who, feeling a vague uneasiness, blame their mood on a failure to meet a personal standard of competence, courage, attractiveness, empathy, nurture, loyalty, or honesty and become vulnerable to the emotion that, in English, we call guilt. An adult might wonder if she had been rude to a friend, told a lie, boasted excessively, harbored a prejudice, or not made enough of her life. In other cultures, she might interpret the same vague uneasiness as meaning that she had offended an ancestor or broken a taboo on eating, sex, or posture with an elder. The list of possible ethical lapses is so long few people in any culture will have trouble finding some violation of a moral standard to explain the unwelcome feeling.

The high-reactive 11-year-olds who described themselves as "feeling bad if one of my parents says I did something wrong" displayed more biological signs of amygdalar activity than either the high-reactives who did not endorse this trait or the low-reactives who also admitted to feeling bad following parental criticism. Further, more mothers of high- than low-reactive girls described them as "sensitive to punishment," and these sensitive girls showed greater right parietal activation than the high-reactive girls whose mothers considered them less sensitive. High-reactive boys described as "sensitive to parental punishment" had larger Wave 5 values than the high- or low-reactive boys who were relatively indifferent to chastisement. Thus, the limbic lability of high-reactives renders them susceptible to changes in feeling tone which, we believe, they interpret as signals of an ethical lapse.

Detection of a semantic or logical inconsistency among one's beliefs, or between one's beliefs and actions, also generates a subtle feeling of uneasiness. Because high-reactive adolescents experience uneasiness more readily and more regularly than others, they are more highly motivated to examine their beliefs in order to ferret out the inconsistencies that create uncertainty. Our high-reactive adolescents were more likely than their peers to complain that they think too much—a salient quality of Jung's introvert.

Temperament and Psychopathology

The combination of intrusive changes in feeling tone and concern with the private evaluations of others places high-reactives at a somewhat higher risk than most for developing serious anxiety over social interactions with non-intimates (Rosenbaum et al., 2000). Clinicians classify such adolescents and adults as *social phobics,* a category with a lifetime prevalence in America and Europe that ranges between 5 and 15 percent (Heiser, Turner, and Beidel, 2003). The defining feature of social phobia is a reluctance to interact with strangers because of worry over what they will think. Most people feel a brief flush of embarrassment if they believe their appearance or action might have provoked a silent criticism, especially from a stranger. Social phobics experience more intense feelings of anxiety, shame, or tension on more occasions and therefore try to avoid situations that might produce these feelings.

Social phobics, who are uncomfortable at parties, on group tours, and in jobs requiring interaction with others, are not afraid of a specific person but of a situation. American phobics fear that the stranger will evaluate them in an undesirable way. The same class of patient in Japan is afraid of disturbing the mental serenity of another (Clarvit et al., 1996). The feature common to the adults in the two cultures is a concern with another's thoughts.

There are two types of social phobics. One type anticipates a critical evaluation from another because they possess undesirable features (for example, a cleft palate, a short stature, or a limited vocabulary). A second type, more temperamental in origin, has inherited a labile physiology and usually experiences a warm face, moist palms, and a tight stomach when interacting with, or anticipating meeting, strangers.

In an earlier study, we found that a larger proportion of adolescents who had been categorized as inhibited rather than uninhibited in the second year had developed social phobia by 13 years of age (Schwartz, Snidman, and Kagan, 1999). Preschool children who were exceptionally inhibited before school entrance were more likely to preserve signs of social anxiety than those who developed an inhibited persona after school entrance (De Wit et al., 1999; Cooper and Eke, 1999).

Fortunately, most high-reactives do not become social phobics (Richards et al., 2002; Heiser, Turner, and Beidel, 2003). A majority find an adaptive niche that protects them from dealing frequently with unfamiliar people on an unpredictable schedule. Many vocational roles permit this protection from uncertainty, while simultaneously awarding respect, intellectual challenge, and financial security. One of our inhibited children told an interviewer that he wanted to be a scientist when he grew up. When the interviewer asked why, the boy paused for about 20 seconds and replied, "I like being alone." This boy is now a graduate student in physics.

Some readers might wonder whether all 4-month-olds should be tested for a high-reactive bias so that their parents could implement special rearing regimens to reduce the risk of later social phobia. Such a proactive strategy has problems. Because no more than one-third of high-reactive infants will develop social phobia, the prediction from infancy will be incorrect 2 out of 3 times. Creating unnecessary apprehension in two-thirds of parents with a high-reactive infant does not seem prudent. A shy, timid persona in an 11-year-old is not an inevitable outcome of a high-reactive profile at 4 months.

Low-reactives face different risks. They are more likely to ignore some community norms on behavior because they are less concerned over adult criticism or the consequences of risky decisions. One of our low-reactive adolescents forged a letter from her parents to the admissions committee of a private school where she had been accepted because she did not want to attend the school. A low-reactive temperament is a good predictor of adult psychiatric problems among North Americans and Europeans who grew up in families where academic accomplishment was not encouraged. Indifference to teacher evaluations, when combined with lax educational standards at home, can increase the risk of academic failure, which in industrialized societies can frustrate vocational success, marital harmony, and a feeling of dignity (Hofstra et al., 2002).

An extremely uninhibited profile in children from economically disadvantaged families represents a modest risk factor for chronic display of physical aggression during adolescence. For example, 6 percent of a large sample of boys from disadvantaged backgrounds were persistently aggressive from their second to their eighth year.

One team of scientists found that an excellent predictor of this asocial trait was failure to show any sign of fear to the unexpected sound of a gorilla while the young child was playing in a laboratory room (Shaw et al., 2003). However, observations from different laboratories affirm that persistent aggression is characteristic of only a small proportion of school-age boys (about 4 percent) who, as adolescents, displayed minimal guilt over ethical violations (Brody et al., 2003; Frick et al., 2003; Colder, Mott, and Berman, 2002). It may not be a coincidence that a higher proportion of criminals than social phobics show minimal activity in the amygdala when they see neutral faces that signal the imminent onset of a painful stimulus (Veit et al., 2002).

Thus, low-reactive boys raised in neighborhoods where peers drift toward delinquency and in families that do not effectively socialize aggressive behavior are probably at slightly higher than average risk for a delinquent career (Farrington, 2000). But low-reactive boys living in nurturant families, free of psychopathology, that effectively socialize aggression do not have higher rates of delinquency. Indeed, these boys are likely to be popular with their peers.

Our data do not support the intuition that low-reactive boys might be at risk for attention deficit hyperactivity disorder (ADHD). Only 2 of our low-reactives received this diagnosis and, except for a very small number of low-reactives who appeared restless and distractible in our laboratory, over 98 percent did not manifest any signs of ADHD. We suspect that valid cases of ADHD originate in a temperament that was not prevalent in our sample.

Only one child among the 237 subjects in our study developed very serious psychopathology. This girl showed low motor activity and frequent crying at 4 months, an average fear score at 14 months, and symptoms of the autistic spectrum at 21 months. When she was 6 years old she was diagnosed—correctly, we believe—with pervasive developmental disorder, and she was not seen at 11 years of age.

Temperament and the Pursuit of Happiness

The egalitarian premise American society celebrates is not completely compatible with the possibility that individuals possess different feeling tones, for consciousness is where "will" and therefore freedom of choice resides. Most citizens who readily accept inherent differences in exceptional mathematical or musical talent, athletic skill, and longevity are less receptive to acknowledging variation in qualities of consciousness, because that possibility implies an obstacle to communication and leads to the uncomfortable conclusion that all are not equally capable of worry, shame, or guilt, as well as joy and empathy for another. These emotions are as fundamental to the harmony of a society as the interbreeding of two animals is to evolution. If a chronically jovial mood is partly the product of an inherited physiology, some people have an unfair advantage, for a sanguine mood ought to be earned. There is injustice if it is simply a gift that nature awards in the beginning. A temperamental bias should not alter the odds of attaining the happiness and personal satisfaction our society values.

The possibility that some adults find it difficult to experience the states that a majority believes are life's primary goals is disturbing to many. Every person should be capable of the same degree of satisfaction if his talent and perseverance have been rewarded. If some who are idle and selfish nevertheless feel happy, while some hard-working empathic persons do not, the principle of fairness has been violated. Most people believe that happiness requires the capture of a desired goal in a particular way. The pleasure extracted from wealth achieved by fraud is not psychologically equivalent to wealth acquired through skill and hard work.

Thus, a resistance to temperamental ideas, though crumbling, is due not only to a desire to deny biological determinism but also to an ethical canon that defines what each individual should do to feel satisfied at the end of each day and, later, as life's end nears. This blend of Puritanism and egalitarianism, common among Americans, is inconsistent with the fact of temperamental variation which, occasionally, denies joy to the prudent, persistent, and talented. A cartoon in *The New Yorker* magazine captures this conundrum: A man is talking to a friend in front of an expensive mansion and three luxury cars. The caption reads: "If I had known happiness was in the genes, I wouldn't have worked so hard."

Temperament and Geography

Biologists estimate that genetic differences between reproductively isolated groups account for only 3 to 5 percent of the variation among human populations (King and Motulsky, 2002). Nevertheless, long-term residents of Asia, Africa, Australia, India, Latin America, Scandinavia, the Balkans, and the Middle East have typical distributions of body types, facial features, pigmentation, blood groups, and disease vulnerabilities that can be traced to very slight differences in their genetic constitution. For example, the proportion of Rh-negative blood types is less than 1 percent in China but greater than 15 percent in Europe (Cavalli-Sforza, 1991). The differences among geographically separated populations are especially obvious when Asians, Africans, and Europeans are compared with one another. But even among European samples, the genomes of adults from Scandinavia vary a little from those living in Spain, Italy, and the Balkans. In general, the greater the geographic distance between two groups and the longer the period of reproductive isolation, the greater the genetic variation.

If, as is likely, groups that have been reproductively isolated for a long time differ in the distribution of inherited neurochemical profiles, as they do for blood types, they should also vary in their distribution of temperaments (Chen et al., 1999). This issue is delicate because of racial and ethnic strife in the world, and it comes as no surprise that many scientists avoid studying genetically based differences in moods or competences among reproductively isolated populations.

The most consistent evidence on geographic variation in temperamental biases comes from comparisons of Asian and European-Caucasian infants and children. Despite a great many individual exceptions, most Europeans and Asians have been reproductively isolated for close to 30,000 years, or about 1,000 generations. To put this time frame in perspective, we know that distinctive behavioral profiles can be created in many animal species after only 20 generations of selective breeding (Mills and Faure, 1991; Plotkin, 1988; Trut, 1999; Belayev, 1979).

Studies from our own and other laboratories reveal that a high-reactive temperament is more common among Caucasian than among Chinese infants. Four-month-old infants born to Chinese parents in Beijing were far less active and less likely to cry than Caucasian infants born in Boston or Dublin (Kagan et al., 1994). Newborn Asian-American infants are calmer, less likely to remove a cloth placed on their face, and more easily consoled than Caucasian-American infants (Freedman and Freedman, 1969).

The greater placidity of Chinese infants was revealed in the reactions of 11-month-olds to a simple incentive. The examiner first covered a toy with a cloth and then allowed the infant to lift the cloth to retrieve the toy. After the infant performed this sequence 4 times, a novel toy appeared when the infant lifted the cloth on the fifth trial. Most of the Caucasian infants showed behavioral signs of surprise, either in face or posture; for example, they became immobile temporarily when they saw the unfamiliar toy. The Chinese infants were less likely to display any observable reaction to the unexpected event (Camras et al., 1998). Further, Japanese infants, compared with Caucasian-Americans, were less easily aroused, less likely to cry in response to an inoculation, and less distressed by restraint of their arms (Caudill and Weinstein, 1969; Lewis, Ramsay, and Kawakami, 1993).

However, when the low-reactive Asian infants grow up, they do not show the attraction to novelty and exuberance characteristic of many low-reactive Caucasian children. Chinese mothers of 6- and 7-year-olds born in Shanghai more often described their children as showing low activity levels and minimal impulsivity compared with the reports of Caucasian mothers living in the Pacific Northwest. The parents of school-age Thai children reported that they were concerned over their children's low energy level, low motivation, and forgetfulness. The parents of Caucasian-American children, by contrast, reported greater concern with aggression, disobedience, and hyperactivity (Weisz et al., 1995, 2003). Thus, we confront the paradox of low-reactive 4-month-old Chinese infants becoming subdued older children, while low-reactive Caucasian infants become exuberant.

Asian and Caucasian populations differ in a DNA segment that defines the promoter region monitoring the production of the serotonin transporter molecule and in alleles that affect the receptors for gastrin-releasing peptide (Marui et al., 2004). This peptide mutes activity in the basolateral nucleus of the amygdala by enhancing GABA activity. Further, clinicians have learned that Asian-American psychiatric patients require a lower dose of psychotropic drugs than Caucasian-American patients with the same symptoms living in the same region, implying that Asians are at a lower level of limbic arousal (Lin et al., 1986). Finally, anthropologists tell us that the ancient Japanese celebrated the feeling of *kami* (translated as surprise), an emotion that might be more salient among those with a normally lower level of cortical and autonomic arousal.

These and other scattered facts invite reflection on two enduring philosophies that have had differential appeal to Asians and Europeans. Reformation Christian theologians, who came from northern rather than southern Europe, emphasized the inherently dysphoric feature of human nature. Both Martin Luther and John Calvin believed that anxiety and guilt were endemic to the human condition. For Calvin—a chronic melancholic—the most desirable psychological state was freedom from anxiety and guilt, but he doubted that he or others could ever escape the continued disquiet of worry and self-reproach.

Buddhist philosophy, more attractive to Asians, makes attainment of a serene mind, rather than absence of guilt, the primary goal to achieve. The Buddhist imperative urges the elimination of all desire for material and sensory pleasures because frustrated wishes are the primary cause of a feeling of suffering that makes serenity impossible. Each person attains this idealized tranquility when acute awareness of the world and of self is obliterated. However, the unpleasant feeling that accompanies the inability to gain a desired goal, or the loss of one that has been gratifying, differs from the anxiety or guilt that accompanies criticism, failure, or self-reproach. Put simply, sadness and worry are distinctly different feeling states. Perhaps that is why English has more words to describe the variation in anxiety (anxious, fearful, worried, concerned, terrified, vexed, and troubled, to name a few) than the variation in sadness.

It is tempting to speculate that temperamental differences between Asians and Europeans made a small, but perhaps real, contribution to the attractiveness of these two philosophies. If large numbers of European adults experienced high levels of cortical and autonomic arousal and interpreted their feelings as guilt, fear, or anxiety, a philosophy that urged serenity of mind would have met resistance because that perfect state would have seemed unattainable. By contrast, a philosophy that accepted anxiety and guilt as definitive of the human condition would seem less valid to those whose temperament permitted many moments of serenity. Achieving a state of mind that freed one from suffering might seem a real possibility to this second group. Perhaps biology, culture, and personal experience came together in a symbiotic alliance to influence the preferred philosophical orientations of these two populations.

Temperament and Genes

Disagreements over the differential contributions of biology and experience to human psychological properties, and especially temperamental biases, have subsided a bit as scientists realized, and should have earlier, that the question was phrased incorrectly. Imagine that two painters had worked together on the same canvas: what sense would it make for a viewer to ask how many square inches each artist had contributed to the finished work? A more useful perspective recognizes the continually collaborative contribution of biology and experience to psychological outcomes.

Exciting advances in genetics over the last few decades have emboldened some behavioral scientists to argue that genetic mechanisms account for a great deal of the variation in many human characteristics. This claim is too strong, for the following reasons. The mathematical equation most often used to estimate the heritability of a human trait assumes that (1) genetic and environmental forces are additive, (2) interactions among genes and between genes and environment are small, and (3) the relation between variation in the relevant gene (or genes) and variation in the trait is linear. Each of these assumptions is vulnerable to critique (Nijhout, 2003). Most biological or behavioral phenotypes are not a function of additive factors but products of nonlinear interactions among genes and between genetic propensities and experiences.

Current heritability estimates for human intelligence and many personality traits may be too large because they fail to evaluate the environmental contribution directly and therefore incorrectly estimate the interaction between genes and environment. For example, the heritability of IQ scores in 7-year-olds is close to zero for children growing up in poverty but relatively high for those growing up in middle-class homes (Turkheimer et al., 2003). Rats who differed in the tendency to avoid electric shock when a signal for the shock occurred were selectively bred over many generations to produce two distinct strains: one that regularly avoided the shock and another that did not. Yet the heritability of these behaviors was less than 0.2 (Brush, 2003). It strains credibility to argue that the heritability of more complex human properties monitored by more genes could be as high as 0.5 or 0.6.

Finally, behavioral geneticists studying human traits consistently report little effect of shared environments. An important reason for this result is that the most significant psychological influences on a person's beliefs about self and others consist of idiosyncratic, symbolic constructions for which scientists have no sensitive methods of measurement. An 11-year-old high-reactive girl who knows that her mother is very shy broods on the possible causes of this trait and constructs beliefs about the malleability of her own timidity. Because these private constructions render each person's psychological environment unique, there may be no shared environmental variance for a number of human traits. The variance that is currently assigned to the environment is, necessarily, unshared.

Temperament and Context

Each person holds representations of the events most likely to occur in particular contexts, whether forest, street, workplace, airport, stadium, home, television screen, or laboratory. The local context, like a channel setting on a television set, primes the brain to expect a particular envelope of most probable experiences. Men holding machine guns are discrepant in most settings, but not on television or cinema screens. The sudden appearance of a clown evokes a reflex immobility, and often a scream of fear, from a child in a laboratory playroom. The sudden appearance of a clown at a circus evokes a smile in the same child because the brain/mind was primed for the event in that setting. A woman who feels the unexpected sensation of a racing heart will interpret this feeling in different ways if she has just finished jogging or is walking alone at night in an unfamiliar city. A feeling of emptiness on rising in the morning could be construed as sadness if a parent had died recently, but as a result of a bad night's sleep if there had been no personal loss.

Adults subjected to both psychological and physical stressors in their home showed a smaller increase in heart rate and blood pressure if their pet dog or cat were nearby than if they were tested alone, with their pet in another part of the house (Allen, Blascovich, and Mendes, 2002). Similarly, adults stressed in a laboratory setting secreted less cortisol if a close friend was with them in the room than if they were tested in isolation (Heinrichs et al., 2003). The same process can be observed in rats. An injection of cocaine into the brain has one consequence if the rat is in a familiar home cage but quite another if the animal is in an unfamiliar place. The unique conditions existing in different laboratories even influence the behavior of genetically identical mice tested with identical procedures on exactly the same apparatus (Crabbe, Wahlsten, and Dudek, 1999). Because every event is automatically evaluated with respect to the situations in which it normally occurs, all events should be conceptualized as *events in a context*.

Questionnaires, homes, laboratory rooms, classrooms, and fMRI scanners are distinctive contexts that award special meanings to sentences intended to interpret the observations collected in each setting. The category *neuroticism,* a popular personality concept derived from a questionnaire, is regarded by psychologists as an inherent personal trait. However, neuroticism scores increased in Germany and Japan following the defeat of these nations in World War II and decreased among the victorious Americans (Lynn and Hampson, 1977). That is why Gottlob Frege argued that complete propositions, which imply a context that selects one meaning from a set of alternatives, and not single words, are the proper units of meaning. The same word should not be used to describe a rat's avoidance of a brightly lit area, conditioned immobility in a mouse, an adolescent's report of nervousness when speaking in front of the class, and a child's reluctance to sleep at a friend's house.

The specific words people choose to describe their emotions to another person usually contain information about the situation that created the feeling and therefore help listeners appreciate the quality of the speaker's emotion. For example, if a speaker says that her first trip to China was fun, a listener knows that the unfamiliar situations were assimilated. If, however, she said the trip was frustrating, the listener knows that many situations were not assimilated as easily. If a speaker confesses to feeling embarrassed after giving a speech, the listener knows that he believed the audience was dissatisfied with the performance. But if he had said that he felt guilty, the listener would infer that the speaker believed his performance did not meet his personal standards, independent of the views of the audience.

The reluctance to limit a psychological property to a class of situations is obvious in essays on morality. Western theorists continue to debate whether fairness, justice, honesty, kindness, and restraint on harming another are a priori moral imperatives. However, much, not all, of this disagreement melts away if we insist that each action be contextualized. It is easy to argue that killing another without provocation is morally wrong. But administering a lethal dose of morphine can be defended on ethical grounds if requested by an elderly cancer patient in chronic pain.

The probability of an adult helping a stranger who appears blind, based on direct observations on city streets, is very high in Rochester, New York, but low in Chattanooga, Tennessee. However, Chattanooga citizens are more likely than those in Rochester to help a stranger who appears injured. Thus, we cannot compare the two cities, nor individuals, on an abstract quality of altruism (Levine, 2003). Once the context of a moral action is specified, it is easier to sort behaviors and intentions into moral and immoral categories. The concepts that will dominate psychology a century from now will include a reference to the context and the agent's expectations for the events in that context. No current concept in psychology includes these theoretically essential features (Mischel, 2004).

It is not clear why American and European social scientists maintain a preference for broad psychological properties for individuals that ignore the contexts in which they act. Some might argue that this habit comes naturally because the human mind desires, and enjoys, logical consistency. It is easy to find a fact inconsistent with a narrow, context-delimited generalization that spoils the illusion of understanding, and harder to discover a fact that requires rejection of a broad conceptualization. The mind wants a story, not an almanac of discrete facts.

The attraction to words for psychological qualities that are free of contextual constraint owes some of its appeal to the fact that many social scientists interested in human behavior—not all—hold philosophical biases that favor particular societal arrangements and human qualities. The truth value of these philosophical positions is often judged by their semantic consistency. By contrast, the criterion for truth among most biologists, chemists, and physicists is correspondence between a statement and the empirical evidence. The semantic coherence of an argument, although always desirable, is less critical. That is why early twentieth-century biologists gave up their prior, pleasing belief in a seamlessly connected brain and accepted the reality of a physical separation between neurons. Experiments indicate that under some conditions electrons behave like particles, while under others they behave like waves. Even though it is semantically inconsistent to state that an electron can be a particle or a wave, physicists accept this conclusion because correspondence with evidence is more critical than semantic coherence.

We have suggested that a low-reactive infant who becomes an uninhibited child could become a popular leader with peers if raised in some contexts, but a criminal if reared in others. The statement "A low-reactive infant/uninhibited child could become either a highly competent, trial lawyer or a prisoner on death row, depending on childhood experiences" has a compromised semantic coherence, but it is in closer correspondence with the facts.

The American psychologists' preference for processes and properties that are free of contextual constraints is due, in part, to philosophical assumptions traceable to the ancient Greeks, but at variance with philosophical positions that were, and to some extent remain, more popular in China (Nisbett et al., 2001). The material world in the European conception consists of individual entities with stable inherent features. Although these entities—objects, animals, people, atoms—enter into relations with others, it is presumed that their fundamental properties remain unchanged because each can be transformed back into its original form. Although the properties of oxygen are altered when it is combined with hydrogen to make water, scientists can split the water molecule and recover the oxygen.

Classic Chinese philosophers, by contrast, regard the relations in which an entity participates as comprising some of its primary features. A *cup* can be filled, emptied, picked up, given, thrown, cleaned, or placed on a table, a shelf, or in a trash can. Similarly, the act of *running* can be displayed by time, water, an animal, or a human, and its meaning depends on the object to which it is linked. The profile of brain activity evoked in a person judging whether a dog can *run* is distinctly different from the profile generated when the person judges whether a human can *run* (Mason, Banfield, and Macrae, 2004). In order to "know" the meaning of a *cup* or *running*, one must specify the context.

Temperament and Cultural Context

The influence of culture, which represents a continued, rather than a transient, context of development, has been lost in the dizzying excitement over the many elegant and significant discoveries in biology. The mood of many scientists, shared by the public, resembles the ambiance a hundred years ago when the hypothesis of the gene—fifty years before Crick and Watson described its structure—pushed aside the influence of local ecology on the selection of traits and led many biologists to announce that all animal evolution could be explained by mutations affecting differential survival. Fortunately, biologists came to their senses by the late 1940s, and evolutionists like Ernst Mayr argued persuasively that natural selection had to be combined with mutation and recombination to explain the changing distribution of the animal features that define a species. The size of hummingbird bills, for example, is affected by the type of available flowers, and the distribution of both bill sizes and types of flowers change over time (Temeles and Kress, 2003). In order to understand changes in the features of a species, one must know its ecological context and its mutual interactions with other life forms.

A culture can have extraordinary influence on the probability of deviant behaviors. During a brief period of anxiety over witches among the Gusii of Kenya (from 1992 to 1994), three men murdered their mothers, by burning, because they were convinced they were witches. None was punished or ostracized for this act of matricide (Ogembo, 2001). A very small number of adults living in Sweden, some of whom were immigrants from other nations, had made a nonlethal suicide attempt (0.5 percent of the sample of over 23,000). In order to predict this rare act, it was necessary to combine the person's gender, national origin, and social class. Two groups were most likely to have made a suicide attempt: lower-class women born and brought up in Finland and upper-middle-class women born and reared in Asia (Westman et al., 2003).

A subdued, quiet, inhibited child was more prevalent than an ebullient one in Puritan New England; these proportions are reversed today. A high-reactive child in Tibet, where a quiet, cautious persona is encouraged and strangers are rare, is less likely to develop social anxiety than a high-reactive child raised in a large city who must deal with strangers holding unknown intentions. Japanese parents teach their children that every person has one style of social behavior when interacting with nonintimates, called *tatemae,* and a quite different style, called *honne,* when interacting with close friends and relatives. However, many American children are taught to "be true to youself"; each person's behaviors and speech ought to be an honest reflection of his intentions. Because American adolescents learn that this standard is frequently violated, interactions with unfamiliar people, whether realized or anticipated, can become a source of uncertainty for those with a high-reactive temperament.

The individual is the primary social entity in Western society. Each person must attain salvation, wealth, status, power, accomplishment, or happiness on her own. Community praise for success, and blame for failure, are placed primarily on the person. By contrast, the imperative for Asian youth is to seek harmony with, and become part of, a group: first family and later peers and community. Each person's pride or shame rests on the successes or failures of the groups of which he is a member, and not only on his personal talent or perseverance (see Weisz et al., 2003).

Although adolescents can give priority to an individualistic or a communal attitude, high- and low-reactives find the two ethics differentially friendly. High-reactive inhibited children feel a bit more secure in a social structure that sets relatively consistent rules for behavior, does not regularly pose demands for excessive risk, and rewards loyalty to the community's standards with praise and status. This category of child is more vulnerable to uncertainty in a competitive society where accomplishment requires entrepreneurial risk, a competitive style, and dealing with strangers on an unpredictable schedule. Low-reactive, uninhibited children are less threatened by an individualistic imperative, enjoy the excitement of risk and meeting new people, and are more likely to bridle in a society where deviance is punished, whether in the form of extreme talent, lack of civility, or domination of others.

Do We Need Two Vocabularies?

Temperamental biases are described with psychological terms. Hence, it is important to ask whether it will ever be possible to translate sentences describing the behavioral and phenomenological properties of temperamental types into sentences that contain only biological words. At present this question cannot be answered, but there are reasons to be skeptical about ever having a faithful translation. The chemist Roald Hoffman reminds us that it is not even possible to translate the chemical description of the oxidation of iron into the vocabulary of physics without losing the central meaning of the "oxidative state of a molecule."

The problem is captured by a frustrating nineteenth-century discussion between Heinrich von Helmholtz, who had written on the physics of tone as the basis for music, and Johannes Brahms, the prolific composer. The two men were unable to understand each other because Brahms spoke of form and counterpoint while the physicist talked about sine waves and spectra. Although the latter are the physical foundations of musical form and counterpoint, Brahms could not conceive of replacing the vocabulary used to describe a symphony with von Helmholtz's concepts.

A description of the patterns of neuronal excitation that occur when a rat enters a cage containing a novel object cannot replace the statement "Rats prefer to enter places where they have encountered unfamiliar objects in the past," even though we want to know what circuits were activated and what chemicals were released when a rat placed his paws in a cage that contained a novel object and ignored the empty cage nearby.

Consider a thought experiment. Constriction of the capillaries of the head leads to a sensation of pain that most members of our culture call a headache. To relieve the pain, we are likely to take aspirin. Members of another culture, however, who assume that they had been bewitched, will search for the malevolent force. Because the cascade of biological events within the skull that produced the pain can lead to different interpretations and actions, knowledge of the brain states does not permit accurate prediction of the psychological reaction of a person whose head hurts.

The need for distinct vocabularies for biological and psychological events has an analog in Bohr's insistence that physicists must use the concepts of classical physics, not those of quantum theory, to describe an experiment even though the latter represent the foundation of the experimental procedures. When Edward Teller challenged this dualist position, Bohr replied that if the experiments were summarized in the language of quantum mechanics the two of them would be imagining their conversation rather than sitting together drinking tea.

The semantic network for a neurobiological term usually differs from the network of what appears to be the same word in a psychological text. The neurobiological network in which the word *fear* is a salient node when the subjects are mice or rats has *amygdala, conditioned freezing,* and *electric shock* as central terms. However, the salient nodes in the psychological network for fear in humans includes *criticism, failure, illness,* and *pollution.* Thomas Kuhn (2000) used the example of the French word *douce* and the English word *sweet* to make this point. Although sugar would be called *douce* by the French and *sweet* by Americans, only the French would use the word *douce* to describe a bland-tasting soup and only Americans would apply the word *sweet* to ingenuous girls. Thus, *douce* and *sweet* have related but not identical meanings.

The behaviors, thoughts, and feelings that define a temperamental bias are the final product of a series of cascades that began with an external stimulus or a spontaneous biological event. The forms that comprise each cascade must be described with different vocabularies. That is, genes, chromosomes, neurons, animals, and species require distinct predicates because each has unique functions. Genes mutate, chromosomes separate, neurons synapse, animals mate, and species evolve. *High-reactive* and *low-reactive* are descriptions of infants—not descriptions of genes, neurons, or circuits.

What Criteria Have Priority?

Economists and evolutionary biologists have an advantage over social scientists because the first two disciplines have achieved a consensus regarding a criterion to use when evaluating one phenomenon as better or more useful than another. Economists use the easily measured variable of cost. Some economists have argued, for example, that because the cost of vaccinating every child against measles is greater than the cost of treating those who come down with the disease, perhaps our society should consider the utility of not requiring universal vaccination.

Evolutionary biologists rely on the concept of inclusive fitness to evaluate the differential significance of an animal's many features. Inclusive fitness is measured by the number and vitality of an individual's offspring and the reproductive success of all genetic relatives, compared with the fecundity of other animals in that ecology. This concept presumably explains the changing distributions of animal species and their features from life's origin 3.5 billion years ago to the present moment. The survival of a species does not imply a correlation with any other desirable psychological property, like sensory acuity, quality of spatial memory, or motor coordination. The fact that large numbers of animals survive over thousands of generations carries with it no ethical implications whatsoever. Gazelles prefer fewer tigers; tigers prefer fewer humans; humans prefer fewer flies. The biologist simply wants to understand why the numbers are the way they are and what genetic and ecological events led to changes in these numbers over time.

By contrast, social scientists cannot agree on a criterion that awards priority to any one of the large number of human qualities that wax and wane over a lifetime, century, or millennium, and none of these candidates can be quantified as easily as cost and fecundity. In nineteenth-century Britain and America, happiness and freedom from coercion were celebrated because of the assumption that pursuit of these goals sustained economic progress. The Japanese during the same historical era valued each person's contribution to the harmony of his family and the community. During a brief interval in the twentieth century, roughly from 1910 to 1960, followers of Freud's ideas argued that lifting the repression on id impulses in order to free libido was the ideal to chase. But the wildness of Western youth in the late 1960s revealed the flaws in this view. More recently, commentators who proposed that long life was the most attractive criterion had to confront the increasing number of bedridden 90-year-olds wishing to die because life had lost all pleasure.

But if not happiness, freedom, a liberated libido, or longevity, what property—or properties—might function as an executive monitor of psychological development? Even casual reflection on this question reveals that no candidate has the ethical neutrality of cost in dollars or number of healthy offspring. Every property nominated—social status, wealth, years of education, number of friends, number of orgasms, children nurtured, charitable works, promises kept, books written, discoveries made, places visited, or days free of anger—hides a value bias and frustrates a desire for a rational answer to what is best. It is not possible to defend the absolute priority of any of these candidates with either rational argument or empirical evidence. Each is a culturally limited, ethical premise.

Ethnically homogeneous villages with little variation in wealth or status (there is always some variation) will idealize different human qualities from those selected by modern, industrialized states with extraordinary variation in privilege and power. The latter communities might reason, with John Donne, that the envy and anger of large numbers of disenfranchised citizens are threats to the comfort of the privileged and an incentive for empathy and, perhaps, a little guilt over the indifference displayed to those who appear to live less happy lives. Some societies might choose to reduce the range of wealth and power, as the People's Republic of China tried to do a half century ago.

Our earlier discussion of feeling tones and the psychological states derived from them provides a possible nominee. Specifically, we propose as a criterion a consciousness minimally perturbed by the dysphoric feeling which accompanies recognition that one has violated a personal standard, or failed to attain a personal goal that one regards as praiseworthy. This proposal can be put more positively: each person wishes to believe that he is acquiring features that match his standards for perfection. This state is attained, always momentarily, by behaving, thinking, or feeling in ways that are in accord with one's understanding of what Plato described as "doing the good."

The behavior of every animal species is monitored by a small number of propensities that have priority. Smells direct the behaviors of dogs; position in the dominance hierarchy influences the actions of chimpanzees. Once children understand the meanings of good and bad—by the third birthday—most of their decisions and behaviors are affected by their evaluation of themselves as good or bad.

Stoic philosophers, like Buddhists, valued a detachment from external goals and obligatory relationships, for the passions surrounding these desires perturb the soul's serenity. Followers of Muhammad attained this state through obedience to the commandments in the Koran. Kant located it in acceptance of the categorical imperative to "act as if the maxim from which you act were to become through your will a universal law." For G. E. Moore, the desired psychological state was attained through intimate relations with others and encounters with beautiful things. Pitirim Sorokin, who had a close brush with death at the hands of the Bolsheviks and eventually became a Harvard sociologist, ended his memoir describing five harsh years of extreme privation by informing readers that what he had learned from witnessing the beginning of the Bolshevik revolution was the necessity of remaining loyal to one's own moral standards.

Striving for this psychological state can, under some social conditions, be inconsistent with the biological imperative to maximize inclusive fitness. The desire to enhance one's perfectability motivates many couples to control their fertility so that one or both can feel the sense of personal accomplishment that comes through education, vocational advancement, travel, artistic pursuits, or a host of other activities. Equally important, the unbridled self-interest linked to maximizing fitness often conflicts with desires to be loyal to, cooperative with, and nurturant toward others who are not kin.

But a rash of books published over the last 25 years claim that unconflicted self-interest is to be expected, given our evolutionary history. After pointing to examples of self-interest in many animal species, these writers imply that because this motive is present throughout nature, humans need not feel ashamed of consistently catering to self. However, anyone with a modest knowledge of the natural world and minimal inferential skill can find examples in nature that support almost any ethical message desired. Those who wish to sanctify marriage can point to pair-bonding gibbons; those who think infidelity is more natural can nominate chimpanzees. To argue that people are naturally sociable, cite baboons; to argue that humans are for the most part solitary, focus on orangutans. If you believe sex should replace fighting, celebrate the Bonobo chimpanzee. If mothers should care for their infants, rhesus monkeys are the model; if fathers should be the primary caretakers, point to titi monkeys. If one feels that women should be in positions of dominance, elephants are the species to emulate. If one feels that men should dominate harems of beautiful women, elephant seals will bolster the case. Nature has enough diversity to fit almost any ethical taste.

Humans are both selfish and generous, aloof and empathic, hateful and loving, dishonest and upright, disloyal and true, cruel and kind, arrogant and humble. But most people feel a little uneasy over an excessive display of the first member in each of those pairs. The resulting dissonance, which is unique to our species, makes us uncomfortable, and we are eager to have it ameliorated. Thus, we suspect that some Americans and Europeans feel better when they read that their unabashed pursuit of pleasure and ego aggrandizement is a natural consequence of their phylogenetic history. The high status of the biological sciences has made it possible for students of evolution to serve as therapists to their community.

The psychological differences between the first humans, who lived in groups, and many contemporary populations are dramatic. Hence, it is useful to ask whether the current psychological profiles of the latter are biologically prepared propensities that have special priority, or dispositions demanded by the social conditions history sculpted over the past twenty thousand years. Most young monkeys in natural settings play with other monkeys. But rhesus monkey infants who had been taken from their mother early in life and placed with an inanimate wire object sit crouched in a corner of a cage away from their peers (Harlow and Harlow, 1966). The capacity to crouch alone in a corner is inherent in the rhesus monkey genome, but actualization of that profile requires very special, unnatural conditions. Thus, it is appropriate to wonder whether the contemporary ethos that urges greater concern with self than with one's primary group is like the monkey's solitary crouched posture because it must overcome a biologically stronger urge to be a loyal, cooperative, and trusting member of a group. Most species that violate their strongest natural propensities risk a loss in inclusive fitness. A warning may be hiding in this biological fact.

One reason for the appeal of the concept of inclusive fitness to explain behavior is its promise to give an unambiguous answer to a question. "Facts" possess different features. Some are pleasing, some are certain, some are useful; and scholars vary in which property they celebrate. It is useful to know that smoking, which many enjoy, increases the probability of developing lung cancer, but that "fact" is neither pleasing nor certain. It is pleasing to learn that adolescents whose parents socialized aggression and encouraged school success are less likely to abuse drugs, but that "fact" is far from certain and, in any case, it is not always useful because democratic governments cannot force parents to behave in prescribed ways. We can say with certainty that a one-second burst of white noise will activate the primary auditory cortex, but neither the utility nor the pleasingness of this fact is obvious.

Most natural scientists prefer certainty over utility and the capacity to please, in part because certainty has an aesthetic appeal. There is beauty in being able to say yes or no to a question. The machines used in scientific inquiry increase the level of certainty because they require well-controlled conditions and permit measurements of components of events that, in their natural state, are coherent wholes. We used machines to measure the biological variables in our 11-year-olds, and we believe that some of the variables we quantified will eventually become part of the definition of some temperaments.

However, an intolerance for ambiguity—a love of certainty— turns investigators' efforts away from quantifying feeling states because no existing procedure or machine can measure them with sensitivity. No fMRI scanner can generate a brain profile that will allow an investigator to predict with confidence whether a person is, at that moment, feeling happy, sad, proud, or guilty. But this frustration, which might be temporary, is not a good reason to ignore the importance of feeling states. Sometimes it is productive to look for the keys where they fell, rather than where the streetlight is brightest.

Coda

We were surprised by the length of the shadows cast by the infant profiles we observed. A 45-minute laboratory observation of 16-

week-old infants revealed dispositions that were preserved in some children for over 10 years. We are also aware of our good fortune in deciding to assess the infants when they were 4 months old. A maturational transition at 2 to 3 months permits 4-month-olds to relate events to acquired schemata, even though they are not mature enough to modulate their behavior when aroused. Had we observed our infants at 1 week or 1 year, we might not have found as persuasive evidence for the high- and low-reactive categories. Nature occasionally opens her gate for a moment to reveal some of the exotic flowers in her courtyard. If the scientist happens to be turned the other way, her sanctum remains a mystery.

REFERENCES

Abercrombie, H. C., N. H. Kalin, M. E. Thurow, M. A. Rosenkranz, and R. J. Davidson. 2003. Cortisol variation in humans affects memory for emotionally laden and neutral information. *Behavioral Neuroscience* 117:505–516.

Adamec, R. E. 1991. Individual differences in temporal lobe sensory processing of threatening stimuli in the cat. *Physiology and Behavior* 49:455–464.

Adamec, R. E., J. Blundell, and A. Collins. 2001. Neural plasticity and stress induced changes in defense in the rat. *Neuroscience and Biobehavioral Reviews* 25:721–744.

Adamec, R. E., and C. Stark-Adamec. 1986. Limbic hyperfunction, limbic epilepsy, and interictal behavior. In B. K. Boane and K. E. Livingston, eds., *The Limbic System*. New York: Raven, pp. 129–145.

Adams, R. B., H. L. Gordon, A. A. Baird, N. Ambady, and R. E. Kleck. 2003. Effects of gaze on amygdala sensitivity to anger and fear faces. *Science* 300:1536.

Adinoff, B., M. D. Devous, S. E. Best, P. Chandler, D. Alexander, K. Payne, T. S. Harris, and M. J. Williams. 2003. Gender differences in limbic responsiveness, by SPECT, following a pharmacologic challenge in healthy subjects. *Neuroimage* 18:697–706.

Allen, K., J. Blascovich, and W. B. Mendes. 2002. Cardiovascular reactivity and the presence of pets, friends, and spouses. *Psychosomatic Medicine* 64:727–739.

Altenmuller, E., K. Schurmann, U. K. Lim, and D. Paulitz. 2002. Hits to the left, flops to the right. *Neuropsychologia* 40:2249–2256.

Amaral, D. G., J. P. Capitanio, M. Jourdain, W. A. Mason, S. P. Mendoza, and M. Prather. 2003. The amygdala: Is it an essential component of the neural network of social cognition? *Neuropsychologia* 41:235–240.

Aniss, A. M., P. S. Sachdev, and K. Chee. 1998. Affective voluntary muscle contractions in startle response to auditory stimuli. *Electromyography and Clinical Neurophysiology* 38:285–293.

Antoniadis, E. A., and R. J. McDonald. 2000. Amygdala, hippocampus, and discriminative conditioning to context. *Behavioural Brain Research* 108:1–9.

Arbelle, S., J. Benjamin, M. Galin, P. Kremer, R. H. Belmaker, and R. P. Ebstein. 2003. Relation of shyness in grade school children to the genotype for the long form of the serotonin transporter promoter region polymorphism. *American Journal of Psychiatry* 160:671–676.

Arcus, D. 1989. Vulnerability and eye color in Disney cartoon characters. In J. S. Reznick, ed., *Perspectives on Behavioral Inhibition*. Chicago: University of Chicago Press, pp. 291–297.

———. 1991. Experiential modification of temperamental bias in inhibited and uninhibited children. PhD diss., Harvard University.

Aston-Jones, G., and F. E. Bloom. 1981. Norepinephrine containing locus ceruleus neurons in behaving rats exhibit pronounced responses to non-noxious environmental stimuli. *Journal of Neuroscience* 1:887–900.

Atchley, R. A., S. S. Ilardi, and A. Enloe. 2003. Hemispheric asymmetry in the processing of emotional content in word meanings. *Brain and Language* 84:105–119.

Auerbach, J., V. Geller, S. Lezer, E. Shinwell, R. H. Belmaker, J. Levin, and R. Ebstein. 1999. Dopamine D4 receptor (D4DR) and serotonin transporter promoter (5-HTTLPR) polymorphisms in the determination of temperament in two-month old infants. *Molecular Psychiatry* 4:369–373.

Baas, J. M. P., C. Grillon, K. B. E. Bocker, A. A. Brack, C. A. Morgan, J. L. Kenemas, and M. N. Verbaten. 2002. Benzodiazepenes have no effect on fear-poteniated startle in humans. *Psychopharmacology* 161:233–247.

Balaban, M. T., N. Snidman, and J. Kagan. Unpublished. Affective startle modulation and electrodermal reactivity in inhibited and uninhibited adolescents.

Barton, R. A., and J. P. Aggleton. 2000. Primate evolution and the amygdala. In J. P. Aggleton, ed., *The Amygdala*. 2nd ed. New York: Oxford University Press, pp. 479–508.

Bassareo, V., M. A. DeLuca, and G. Di Chiara. 2002. Differential expression of motivational stimulus properties by dopamine in nucleus accumbens shell versus core and prefrontal cortex. *Journal of Neuroscience* 22:4709–4719.

Bates, J. E. 1989. Concepts and measures of temperament. In G. A. Kohnstamm, J. E. Bates, and M. K. Rothbart, eds. *Temperament in Childhood*. New York: Wiley, pp. 3–26.

Baumgarten, H. G. 1993. Control of vigilance and behavior by ascending serotinergic systems. In S. Zschocke and E. J. Speckmann, eds., *Basic Mechanisms of the EEG*. Boston: Birkhauser, pp. 231–268.

Baving, L., M. Laucht, and M. H. Schmidt. 2002. Frontal brain activation in anxious school children. *Journal of Child Psychology and Psychiatry* 43:265–271.

Beach, F. A. 1950. The snark was a boojum. *American Psychologist* 5:115–124.

Belayev, D. K. 1979. Destabilizing selection as a factor in domestication. *Journal of Heredity* 70:301–308.

Berns, G. S., S. M. McClure, G. Pagnoni, and P. R. Montague. 2001. Predictability modulates human brain reponse to reward. *Journal of Neuroscience* 21:2793–2798.

Bevins, R. A. 2001. Novelty seeking and reward. *Current Directions in Psychological Science* 10:189–193.

Bevins, R. A., J. Besheer, M. I. Tapalmatier, H. C. Jensen, K. S. Pickett, and S. Eurek. 2002. Novel object place conditioning. *Behavioral Brain Research* 129:41–50.

Bindra, D. 1959. Stimulus change, reactions to novelty, and response deprivation. *Psychological Review* 66:96–103.

Bishop, G., S. H. Spence, and C. McDonald. 2003. Can parents and teachers provide a reliable and valid report of behavioral inhibition? *Child Development* 74:1899–1917.

Blizard, D. A., and N. Adams. 2002. The Maudsley reactive and non-reactive strains. *Behavior Genetics* 32:277–299.

Blumberg, M. S., and G. Sokoloff. 2001. Do infant rats cry? *Psychological Review* 108:83–95.

Blumenthal, T. D. 2001. Extroversion attention and startle response reactivity. *Personality and Individual Differences* 30:495–503.

Bohr, N. 1933. Light and life. *Nature* 131:421–459.

Bornhovd, K., M. Quante, V. Glauche, B. Bromm, C. Weiller, and C. Buchel. 2002. Painful stimuli evoke different stimulus-response functions in the amygdala, prefrontal, insula, and somatosensory cortex. *Brain* 125:1326–1336.

Boysen, S. T., and G. G. Bernston. 1989. Conspecific recognition in the chimpanzee (Pan troglodytes). *Journal of Comparative Psychology* 103:215–220.

Bradley, M. M., B. N. Cuthbert, and P. J. Lang. 1996. Picture media and emotion. *Psychophysiology* 33:662–670.

———. 1999. Affect and the startle reflex. In M. E. Dawson, A. M. Schell, and A. N. Bohmelt, eds., *Startle Modification*. New York: Cambridge University Press, pp. 157–183.

Bradley, M. M., and P. J. Lang. 2000. Affective reactions to acoustic stimuli. *Psychophysiology* 37:204–215.

Brainard, M. S., and A. J. Doupe. 2002. What song birds teach us about learning. *Nature* 417:351–358.

Brandao, M. L., S. H. Cardoso, L. C. Melo, V. Motta, and N. C. Coimbra. 1994. Neural substrate of defensive behavior in the midbrain tectum. *Neuroscience and Biobehavioral Reviews* 18:339–346.

Brandao, M. L., N. C. Coimbra, and M. Y. Osaki. 2001. Changes in the auditory-evoked potentials induced by fear-evoking stimulation. *Physiology and Behavior* 72:365–372.

Brandao, M. L., A. C. Troncoso, M. A. S. Silver, and J. H. Huston. 2003. The relevance of neuronal substrates of defense in the midbrain tectum to anxiety and stress. *European Journal of Pharmacology* 463:225–233.

Brody, L. M., R. E. Tremblay, B. Brane, D. Fergusson, J. L. Horwood, R. Laird, T. E. Moffit, D. S. Nagio, J. E. Bates, K. A. Dodge, R. Loeber, D. R. Lynam, G. S. Petit, and F. Vitaro. 2003. Developmental trajectories of childhood disruptive behaviors and adolescent delinquency. *Developmental Psychology* 39:222–245.

Broman-Fulks, J. J., M. E. Berman, B. A. Rabian, and M. J. Webster. 2004. Effects of aerobic exercise on anxiety sensitivity. *Behaviour Research and Therapy* 42:125–136.

Bronson, G. W. 1970. Fear of visual novelty. *Developmental Psychology* 2:33–40.

Brown, J. S., H. I. Kalish, and I. E. Farber. 1951. Conditioned fear as revealed by magnitude of startle response to an auditory stimulus. *Journal of Experimental Psychology* 41:317–328.

Bruce, J., E. P. Davis, and M. R. Gunnar. 2002. Individual differences in children's cortisol response to the beginning of a new school year. *Psychoneuroendocrinology* 27:635–650.

Brunzell, D. H., and J. J. Kim. 2001. Fear conditioning to tone, but not to context, is attentuated by lesions of the insular cortex and posterior extension of the intralaminar complex in rats. *Behavioral Neuroscience* 115:365–375.

Brush, F. R. 2003. Selection for differences in avoidance learning. *Behavior Genetics* 33:677–696.

Bulbena, A., R. Martin-Santos, M. Porta, J. C. Duro, J. Gago, J. Sangorrin, and M. Gratacos. 1996. Somatotype in panic patients. *Anxiety* 2:80–85.

Bullock, W. A., and K. Gilliland. 1993. Eysenck's arousal theory of introversion-extraversion. *Journal of Personality and Social Psychology* 64:113–123.

Buss, A. H., and R. Plomin. 1984. *Temperament.* Hillsdale, NJ: Erlbaum.

Buss, K. A., J. R. Malmstadt, I. D. Schumacher, N. H. Kalin, H. H. Goldsmith, and R. J. Davidson. 2003. Right frontal brain activity, cortisol, and withdrawal behavior in six month old infants. *Behavioral Neuroscience* 117:11–20.

Buus, S., M. Florentine, and T. Poulsen. 1997. Temporal integration of loudness, loudness discrimination and the form of the loudness function. *Journal of the Acoustical Society of America* 101:669–680.

Byrne, J., and S. J. Suomi. 2002. Cortisol reactivity and its relation to home cage behavior and personality ratings in tufted capuchin (Cebux apella) juveniles from birth to six years of age. *Psychoneuroendocrinology* 27:139–154.

Cahill, L. 2000. Modulation of long term memory storage in humans by emotional arousal. In J. P. Aggleton, ed., *The Amygdala.* 2nd ed. New York: Oxford University Press, pp. 425–441.

Cahill, L., and J. L. McGaugh. 1990. Amygdaloid complex lesions differentially affect retention of tasks using appetitive and aversive reinforcement. *Behavioral Neuroscience* 104:532–543.

Cain, M. E., B. S. Kapp, and C. B. Puryear. 2002. The contribution of the amygdala to conditioned thalamic arousal. *Journal of Neuroscience* 22:11026–11034.

Calkins, S. D., N. A. Fox, and T. R. Marshall. 1996. Behavioral and physiological antecedents of inhibited and uninhibited behavior. *Child Development* 67:523–540.

Cameron, O. G. 2001. Interoception. *Psychosomatic Medicine* 63(5):697–710.

Cameron, O. G., and S. Minoshima. 2002. Regional brain activation due to pharmacologically induced adrenergic interoceptive stimulation in humans. *Psychosomatic Medicine* 64:851–861.

Camras, L. A., H. Oster, J. Campos, R. Campos, T. Ujiie, K. Miyake, L. Wang, and Z. Meng. 1998. Production of emotional facial expressions in European-American, Japanese, and Chinese infants. *Developmental Psychology* 34:616–628.

Camras, L. A., T. Vjiie, K. Miyake, L. Wang, Z. Mang, S. Dharamsi, H. Oster, J. Cruz, A. Merdock, and J. Campos. 2002. Observing emotion in infants. *Emotion* 2:179–193.

Carels, R. A., J. A. Blumenthal, and A. Sherwood. 2000. Emotional responsivity during daily life. *International Journal of Psychophysiology* 36:25–33.

Carlson, J. N., L. W. Fitzgerald, R. W. Keller, and S. D. Glick. 1993. Lateralized changes in prefrontal cortical dopamine activity induced by controllable and uncontrollable stress in the rat. *Brain Research* 630:178–187.

Caspi, A., G. H. Elder, and D. J. Bem. 1988. Moving away from the world. *Developmental Psychology* 24:824–831.

Caspi, A., and P. A. Silva. 1995. Temperamental qualities at age three predict personality traits in young adulthood. *Child Development* 66:486–498.

Caudill, W., and H. Weinstein. 1969. Maternal care and infant behavior in Japan and America. *Psychiatry* 32:12–43.

Cavalli-Sforza, L. L. 1991. Genes, people, and language. *Scientific American* 265:104–111.

Cavigelli, S. A., and M. K. McClintock. 2003. Fear of novelty in infant rats predicts adult corticosterone dynamics and an early death. *Pro-*

ceedings of the National Academy of the United States of America 100:16131–16136.

Cecchi, M., H. Khoshbouei, M. Javors, and D. A. Morilak. 2002. Modulatory effects of norepinephrine in the lateral bed nucleus of the stria terminalis on behavioral and neuroendocrine responses to acute stress. *Neuroscience* 112:13–21.

Ceponiene, R., T. Lepisto, M. Soininen, E. Aronen, P. Alku, and R. Naatanen. 2004. Event related potentials associated with sound discrimination versus novelty detection in children. *Psychophysiology* 41:130–141.

Charlesworth, W. R. 1969. The role of surprise in cognitive development. In D. Elkind and J. H. Flavell, eds., *Studies in Cognitive Development*. New York: Oxford University Press, pp. 257–313.

Chavira, D. A., M. B. Stein, and V. L. Malcarne. 2002. Scrutinizing the relationship between shyness and social phobia. *Journal of Anxiety Disorders* 16:585–598.

Chen, C., M. Burton, E. Greenberger, and J. Dimitrieva. 1999. Population migration and the variation of dopamine D4 receptor (DRD4) allele frequencies around the globe. *Evolution and Human Behavior* 20:309–324.

Cheng, D. T., D. C. Knight, C. N. Smith, E. A. Stein, and F. J. Helmstetter. 2003. Functional fMRI of human amygdalar activity during Pavlovian fear conditioning. *Behavioral Neuroscience* 117:3–10.

Chiappa, K. H. 1997. *Evoked Potentials in Clinical Disorders*. 3rd ed. Philadelphia, PA: Lippincott.

Chotai, J., E. S. Renberg, G. Kullgren, and M. Asberg. 2000. Seasons of birth variations in dimensions of functioning evaluated by the diagnostic interview for borderline patients. *Neuropsychobiology* 41:132–138.

Clarke, C. R., A. C. MacFarlane, D. L. Webber, and M. Battersby. 1996. Enlarged frontal P300s to stimulus changes in panic disorder. *Biological Psychiatry* 39:845–856.

Clarvit, S. R., F. R. Schneier, and M. R. Liebowitz. 1996. The offensive subtype of Taijin Kyofu-sho in New York City. *Journal of Clinical Psychiatry* 57:523–527.

Cobos, P., M. Sanchez, C. Garcia, V. M. Nieves, and J. Vila. 2002. Revisiting the James verus Cannon debate on emotion. *Biological Psychology* 61:251–269.

Codispoti, M., M. Bradley, and P. J. Lang. 2001. Affective reactions to briefly presented pictures. *Psychophysiology* 38:474–478.

Cohen, H., J. Benjamin, A. B. Geva, M. A. Mator, Z. Kaplan, and M. Kotler. 2001. Autonomic dysregulation in panic disorder and in post-traumatic stress disorder. *Psychiatry Research* 96:1–13.

Cohen, R. A., and J. D. Coffman. 1991. Beta adrenergic vasodilator mechanism in the finger. *Circulation Research* 49:1196–1201.

Colder, C. R., J. A. Mott, and A. S. Berman. 2002. The interaction of sex of infant, activity level, and fear on growth trajectories of early childhood behavior problems. *Development and Psychopathology* 14:1–23.

Coleman-Mesches, K., and J. L. McGaugh. 1995. Differential effects of pretraining inactivation of the right or left amygdala on retention of inhibitory avoidance training. *Behavioral Neuroscience* 109:642–647.

Collaer, M. L., and M. Hines. 1995. Human behavioral sex differences. *Psychological Bulletin* 118:55–107.

Comings, D. E., and K. Blum. 2000. Reward deficiency syndrome. *Progress in Brain Research* 126:325–341.

Cooper, P. J., and M. Eke. 1999. Childhood shyness and maternal social phobia. *British Journal of Psychiatry* 174:439–443.

Coplan, R. J., B. Coleman, and K. H. Rubin. 1998. Shyness and little boy blue. *Developmental Psychobiology* 32:37–44.

Cote, S., R. E. Tremblay, D. Nagin, M. Zoccolillo, and F. Vitaro. 2002. The development of impulsivity, fearfulness, and helpfulness during childhood: patterns of consistency and change in the trajectories of boys and girls. *Journal of Child Psychology and Psychiatry* 43:609–618.

Courchesne, E. 1978. Neurophysiological correlates of cognitive development. *Electroencephalography and Clinical Neurophysiology* 45:468–482.

Cox, B. J., L. A. McWilliams, I. P. Clara, and M. B. Stein. 2003. The structure of feared situations in the nationally representative sample. *Journal of Anxiety Disorders* 17:89–101.

Crabbe, J. C., D. Wahlsten, and B. C. Dudek. 1999. Genetics of mouse behavior. *Science* 284:1670–1672.

Cromwell, H. L., and W. Schultz. 2003. Effects of expectations for different reward magnitudes on neural activity in primate striatum. *Journal of Neurophysiology* 89:2823–2838.

Crozier, W. R., and D. Russell. 1992. Blushing, embarrassability, and self-consciousness. *British Journal of Social Psychology* 31:343–349.

Damasio, A.R. 1994. *Descartes' error.* NY: Putnam.

Damasio, A. R., R. Adolphs, and H. Damasio. 2003. The contributions of the lesion method to the functional neuroanatomy of emotion. In R. J. Davidson, K. R. Scherer, and H. H. Goldsmith, eds., *Handbook of Affective Science.* New York: Oxford University Press, pp. 66–92.

Davidson, R. J. 2003a. Affective neuroscience and psychophysiology. *Psychophysiology* 40:655–665.

––––––. 2003b. Right frontal brain activity, cortisol, and withdrawal behavior in six month old infants. *Behavioral Neuroscience* 117:11–20.

Davidson, R. J., and N. A. Fox. 1982. Asymmetric brain activity discriminates between positive and negative affective stimuli in human infants. *Science* 218:1235–1237.

––––––. 1989. Frontal brain asymmetry predicts infants' response to maternal separation. *Journal of Abnormal Psychology* 98:127–131.

Davidson, R. J., D. L. Jackson, and N. H. Kalin. 2000. Emotion, plasticity, context, and regulation. *Psychological Bulletin* 126:890–909.

Davidson, R. J., J. R. Marshall, A. J. Tomarken, and J. B. Henriques. 2000. While a phobic waits. *Biological Psychiatry* 47:85–95.

Davis, M. 1994. The mammalian startle response. In R. C. Eaton, ed., *The Neural Mechanisms of Startle Behavior.* New York: Plenum, pp. 287–351.

Deacon, D., A. Dynowska, W. Ritter, and J. Grose-Fifer. 2004. Repetition and semantic priming of non-words. *Psychophysiology* 41:60–74.

Dean, P., P. Redgrave, and G. W. Westby. 1989. Event or emergency? *Trends in Neuroscience* 12:137–147.

De Bellis, M. D., B. J. Casey, R. E. Dahl, B. Birmaher, D. E. Williamson, K. M. Thomas, D. A. Axelson, K. Frustaci, A. M. Boring, M. Hall, and N. D. Ryan. 2000. A pilot study of amygdala volumes in pediatric generalized anxiety disorder. *Biological Psychiatry* 48:51–57.

De Bellis, M. D., M. S. Keshaven, S. R. Beers, J. Hall, K. Frustaci, A. Masalehdan, J. Noll, and A. M. Boring. 2001. Sex differences in brain maturation in childhood and adolescence. *Cerebral Cortex* 11:552–557.

De France, J. F., S. Sands, F. C. Schweitzer, L. Ginsberg, and J. C. Sharma. 1997. Age related changes in cognitive ERPs of attention. *Brain Topography* 9:283–293.

De Wit, D. J., A. Ogborn, D. R. Offord, and K. MacDonald. 1999. Antecedents of the risk of recovery from DSM III-R Social Phobia. *Psychological Medicine* 29:569–582.

Diefenbach, G. J., M. E. McCarthy-Larzeler, D. A. Williamson, A. Mathews, G. M. Manguno-Mire, and B. G. Bentz. 2001. Anxiety, depression, and the content of worries. *Depression and Anxiety* 14:247–250.

DiLalla, L. F., J. Kagan, and J. S. Reznick. 1994. Genetic etiology of behavioral inhibition among two year old children. *Infant Behavior and Development* 17:401–408.

Dolan, R. J. 2000a. Functional neuroimaging of the amygdala during emotional processing and learning. In J. P. Aggleton, ed., *The Amygdala*. 2nd ed. New York: Oxford University Press, pp. 631–654.

———. 2000b. Emotional processing in the human brain revealed through functional neuroimaging. In M. Gazzaniga, ed., *New Cognitive Neurosciences*. 2nd ed. Cambridge, MA: MIT Press, pp. 1135–1161.

Donzella, B., M. R. Gunnar, W. K. Krueger, and J. Alwin. 2000. Cortisol and vagal tone responses to competitive challenge in preschoolers. *Developmental Psychobiology* 37:209–220.

Doussard-Roosevelt, J. A., L. A. Montgomery, and S. W. Porges. 2003. Short-term stability of physiological measures in kindergarten children. *Developmental Psychobiology* 43:230–242.

Dworkin, R. H., B. A. Cornblatt, R. Friedmann, L. M. Kaplansky, J. A. Lewis, A. Rinaldi, C. Shilliday, and L. Erlenmeyer-Kimling. 1993. Childhood precursors of affective versus social deficits in adolescents at risk for schizophrenia. *Schizophrenia Bulletin* 19:563–577.

Ehrlichman, H., S. Brown, J. Zhou, and S. Warrenburg. 1995. Startle reflex during exposure to pleasant and unpleasant odors. *Psychophysiology* 32:150–154.

Ekman, P., R. J. Davidson, and W. V. Friesen. 1990. Duchenne's smile. *Journal of Personality and Social Psychology* 58:342–352.

Emde, R. N., and J. K. Hewitt, eds. 2001. *Infancy to Early Childhood*. New York: Oxford University Press.

Esteban, A. 1999. A neurophysiological approach to brain stem reflexes. *Neurophysiologieclinique* 29:7–38.

Fairbanks, L. A. 2001. Individual differences in response to a stranger. *Journal of Comparative Psychology* 115:22–28.

Farrington, D. P. 2000. Psychosocial predictors of adult anti-social personality and adult convictions. *Behavioral Science and Law* 18(5):605–622.

Federmeier, K. P., and M. Kutas. 2002. Picture the difference. *Neuropsychologia* 40:730–747.

Fendt, M., T. Enders, and R. Apfelbach. 2003. Temporary inactivation of the bed nucleus of the stria terminalis but not of the amygdala blocks freezing induced by trimethylthiazoline, a component of fox feces. *Journal of Neuroscience* 23:23–28.

Fias, W., P. Dupont, B. Reynvoet, and G. A. Orban. 2002. The quantitative nature of a visual task differentiates between ventral and dorsal streams. *Journal of Cognitive Neuroscience* 14:646–658.

Field, T., M. Diego, M. Hernandez-Reif, S. Schanberg, and C. Kuhn. 2002. Relative right vs. left frontal EEG in neonates. *Developmental Psychobiology* 41:147–155.

Filion, D. L., M. E. Dawson, and A. M. Schell. 1998. The psychological significance of human startle eye blink modification. *Biological Psychology* 47:1–43.

Fox, N. A. 1989. Psychophysiological correlates of emotional reactivity during the first year of life. *Developmental Psychology* 25:264–372.

Fox, N. A., and M. A. Bell. 1990. Electrophysiological indexes of frontal lobe development. *Annals of the New York Academy of Sciences* 608:677–698.

Fox, N. A., S. D. Calkins, and M. A. Bell. 1994. Neural plasticity in development in the first two years of life. *Development and Psychopathology* 6:677–696.

Fox, N. A., H. A. Henderson, K. H. Rubin, S. D. Calkins, and L. A. Schmidt. 2001. Continuity and discontinuity of behavioral inhibition and exuberance. *Child Development* 72:1–21.

Fox, N. A., K. H. Rubin, S. D. Calkins, J. R. Marshall, R. J. Coplan, S. W. Porges, J. M. Long, and S. Stewart. 1995. Frontal activation asymmetry and social competence at 4 years of age. *Child Development* 60:1770–1784.

Frazer, J. G. 1933, 1934, 1936. *The Fear of the Dead in Primitive Religion*, Vols. 1, 2, 3. London: Macmillan.

Freedman, D. G., and N. Freedman. 1969. Behavioral differences between Chinese-American and American newborns. *Nature* 24:12–27.

Freud, S. 1926/1948. *Inhibition Symptoms and Anxiety*. London: Hogarth.

———. 1950. *The Standard Edition of the Collected Psychological Works* of *Sigmund Freud*. Translated by J. Strachey. 24 vols. London: Hogarth.

Frick, P. J., A. H. Cornell, S. D. Bodin, H. E. Dene, C. T. Barry, and B. R. Loney. 2003. Callous-unemotional traits in developmental pathways to severe conduct problems. *Developmental Psychology* 39:246–260.

Frick, K. M., and J. E. Gresack. 2003. Sex differences in the behavioral responses to spatial and object novelty in adult C57BL/6 mice. *Behavioral Neuroscience* 117:1283–1291.

Fridlund, A. J., M. E. Hatfield, G. L. Cottam, and C. Fowler. 1986. Anxiety and striate muscle activation. *Journal of Abnormal Psychology* 95:228–236.

Fried, I., C. L. Wilson, J. W. Morrow, K. A. Cameron, E. D. Behnke, L. C. Ackerson, and N. T. Maidment. 2001. Increased domapine release in the human amydala during performance of cognitive tasks. *Nature Neuroscience* 4:201–206.

Fudge, J. L., K. Kunishio, P. Walsh, C. Richard, and S. N. Haber. 2002. Amygdaloid projections to ventromedial striatral subterritories in the primate. *Neuroscience* 110:257–275.

Funayama, E. S., C. Grillon, M. Davis, and E. A. Phelps. 2001. A double dissociation in the affective modulation of startle in humans. *Journal of Cognitive Neuroscience* 13:721–729.

Gale, A., J. Edwards, P. Morris, R. Moore, and D. Forrester. 2001. Extraversion-introversion, neuroticism-stability and EEG indicators of positive and negative empathic mood. *Personality and Individual Differences* 30:449–461.

Gallagher, M., and P. Holland. 1994. The amygdala complex. *Proceedings of the National Academy of Sciences of the USA* 91:1171–1176.

Gallo, L. C., and K. A. Matthews. 2003. Understanding the association between socioeconomic status and physical health. *Psychological Bulletin* 129:10–51.

Ganis, G., and M. Kutas. 2003. An electrophysiological study of scene effects on object identification. *Cognitive Brain Research* 16:123–144.

Garcia-Coll, C., J. Kagan, and J. S. Reznick. 1984. Behavioral inhibition in children. *Child Development* 55:1005–1009.

Geisler, M. W., and C. Murphy. 2000. Event-related brain potentials to attended and ignored olfactory and trigeminal stimuli. *International Journal of Psychophysiology* 37:309–315.

Gerlach, A. L., F. H. Wilhelm, and W. T. Roth. 2003. Embarrassment and social phobia. *Journal of Anxiety Disorders* 17:197–210.

Goldberg, E., K. Podell, and M. Lovell. 1994. Lateralization of frontal lobe functions and cognitive novelty. *Journal of Neuropsychiatry and Clinical Neurosciences* 6:371–378.

Goldsmith, H. H., and J. J. Campos. 1986. Fundamental issues in the study of early temperament. In M. E. Lamb, A. C. Brown, and B. Rogoff, eds., *Advances in Developmental Psychology*, Vol. 4. Hillsdale, NJ: Erlbaum, pp. 231–283.

Goldstein, A., K. M. Spencer, and E. Donchin. 2002. The influence of stimulus deviance and novelty on the P300 and novelty P3. *Psychophysiology* 39:781–790.

Goodmurphy, C. W., and O. Wik. 1999. Morphological study of two human facial muscles: Orbicularis oculi and corrugator supercilli. *Clinical Anatomy* 12:1–11.

Goodwin, B. 1994. *How The Leopard Changed Its Spots*. New York: Scribner.

Gortmaker, S. L., J. Kagan, A. Caspi, and P. A. Silva. 1997. Daylength during pregnancy and shyness in children. *Developmental Psychobiology* 31:107–114.

Green, E. C. 1996. Purity, pollution, and the invisible snake in South Africa. *Medical Anthropology* 17:83–100.

Greenberg, N., M. Scott, and D. Crews. 1984. Role of the amygdala in the reproductive and aggressive behavior of the lizard, Anolis carolinensis. *Physiology and Behavior* 32(1):147–151.

Grillon, C., R. Ameli, S. W. Woods, K. Merikangas, and M. Davis. 1991. Fear potentiated startle in humans. *Psychophysiology* 28:588–595.

Grillon, C., C. A. Morgan, M. Davis, and S. M. Southwick. 1998. Effects of experimental context and explicit threat cues on acoustic startle in Vietnam veterans with post-traumatic stress disorder. *Biological Psychiatry* 44:1027–1036.

Grossi, G., A. Perski, U. Lundberg, and J. Soares. 2001. Associations between financial strain and the diurnal salivary cortisol secretion of long-term unemployed individuals. *Integrative Physiological and Behavioral Science* 36:205–219.

Gunnar, M. R. 1994. Psychoendocrine studies of temperament and stress in early childhood: Expanding current models. In J. Bates and T. Wachs, eds., *Temperament*. Washington, D.C.: The American Psychological Association, pp. 175–198.

Gur, R. C., F. Gunning-Dixon, W. B. Bilker, and R. E. Gur. 2002. Sex differences in temporal-limbic and frontal brain volumes of healthy adults. *Cerebral Cortex* 12:998–1003.

Habib, R., A. R. McIntosh, M. A. Wheeler, and E. Tulving. 2003. Memory encoding and hippocampally-based novelty/familiarity distinctive networks. *Neuropsychologia* 41:271–279.

Hagemann, D., E. Naumann, J. F. Thayer, and D. Bartussek. 2002. Does resting EEG asymmetry reflect a trait? *Journal of Personality and Social Psychology* 82:619–641.

Halgren, E. 1992. Emotional neurophysiology of the amygdala within the context of human cognition. In J. P. Aggleton, ed., *The Amygdala*. New York: Wiley, pp. 191–228.

———. 2002. Cognitive electrophysiology of human parahippocampal structures. In M. Whitter and F. Wouterlood (Eds.). *The parahippocampal region*. New York: Oxford, pp. 185–237.

Halgren, E., R. P. Dhond, N. Christensen, C. Van Petten, K. Marinkovic, J. D. Lewine, and A. M. Dale. 2002. N400-like magnetoencephalography responses modulated by semantic context, word frequency, and lexical class in sentences. *Neuroimage* 17: 1101–1116.

Hariri, A. R., V. S. Mattay, A. Tessitore, F. Fera, W. G. Smith, and D. R. Weinberger. 2002. Dextroamphetamine modulates the response of the human amygdala. *Neuropsychopharmacology* 27:1036–1040.

Harlow, H. F., and M. K. Harlow. 1966. Learning to love. *American Scientist* 54:244–272.

Hart, D., V. Hofmann, W. Edelstein, and M. Keller. 1997. The relation of childhood personality types to adolescent behavior and development. *Developmental Psychology* 33:195–205.

Hawk, L. W., and A. D. Kowmas. 2003. Affective modulation and prepulse inhibition of startle among undergraduates high and low in behavioral inhibition and approach. *Psychophysiology* 40: 131–138.

Hayssen, V. 1997. Effects of the nonagouti coat-color allele on behavior of deer mice (Peromyscus maniculatus). *Journal of Comparative Psychology* 111:419–423.

Hebb, D. O. 1946. On the nature of fear. *Psychological Review* 53:259–276.

Heinrichs, M., T. Baumgartner, C. Kirschbaum, and U. Ehlert. 2003. Social support and oxytocin interact to suppress cortisol and subjective responses to psychosocial stress. *Biological Psychiatry* 54:1389–1398.

Heiser, N. A., S. M. Turner, and D. C. Beidel. 2003. Shyness. *Behavioural Research and Therapy* 41:209–221.

Heller, W. 1990. The neuropsychology of emotion. In N. C. Stein, B. Leventhal, and T. Trabasso, eds., *Psychological and Biological Approaches to Emotion.* Hillsdale, NJ: Erlbaum, pp. 167–211.

Henderson, H. A., T. J. Marshall, N. A. Fox, and K. H. Rubin. 2002. Psychophysiological and behavioral evidence for varying forms and functions of non-social behavior in preschoolers. Department of Human Development, University of Maryland, unpublished.

Herbener, E. S., J. Kagan, and N. Cohen. 1989. Shyness and olfactory threshold. *Personality and Individual Differences* 10:1159–1163.

Hinton, D., and S. Hinton. 2002. Panic disorder, somatization, and the new cross-cultural psychiatry. *Culture, Medicine, and Psychiatry* 26:155–178.

Hofstra, M. B., J. Van der Ende, and F. C. Verhulst. 2002. Child and adolescent problems predict DSM-IV disorders in adulthood. *Journal of the American Academy of Child and Adolescent Psychiatry* 41:182–189.

Holland, P. C., and M. Gallagher. 1999. Amygdala circuitry in attentional and representational processes. *Trends in Cognitive Sciences* 3:65–73.

Holstege, G., J. J. Van Ham, and J. Tan. 1986. Afferent projections to the orbicularis oculi motoneuronal cell group. *Brain Research* 374:306–320.

Hood, J. D., J. P. Poole, and L. Freedman. 1976. Eye color and susceptibility to TTS. *Journal of Acoustical Society of America* 59:706–707.

Huffman, L. C., Y. E. Bryan, F. A. Pedersen, B. M. Lester, J. D. Newman, and R. Del Carmen. 1994. Infant cry acoustics and maternal ratings of temperament. *Infant Behavior and Development* 17:45–53.

Hupka, R. B., Z. Zaleski, J. Otto, and L. Reidl. 1997. The colors of anger, envy, fear, and jealousy. *Journal of Cross-Cultural Psychology* 28:156–171.

Indovina, I., and J. N. Senes. 2001. Combined visual attention and finger movement effects on human brain representations. *Experimental Brain Research* 140:265–279.

Insel, T. R., Z. Wang, and C. Ferris. 1994. Patterns of vasopressin receptor distribution associated with social organization in monogamous and non-monogamous microtine rodents. *Journal of Neuroscience* 14:5381–5392.

Itier, R. J., and M. J. Taylor. 2004. N170 or N1? Spatiotemporal differences between object and face processing using ERPs. *Cerebral Cortex* 17:132–142.

Jackson, D. C., J. R. Malmstadt, C. L. Larson, and R. J. Davidson. 2000. Suppression and enhancement of emotional responses to unpleasant pictures. *Psychophysiology* 37:515–522.

Jessen, F., C. Manka, L. Scheef, D. O. Granath, H. H. Schild, and R. Heun. 2002. Novelty detection and repetition suppression in the passive picture viewing task. *Human Brain Mapping* 17:230–236.

John, E. M. 1941. A study of the effects of evacuation and air raids on children of pre-school age. *British Journal of Educational Psychology* 11:173–182.

Johnstone, S. J., and R. J. Barry. 1999. An investigation of the event-related slow wave potential (0.01 to 2.0 Hz) in normal children. *International Journal of Psychophysiology* 32:15–34.

Jonsson, P., and M. Sonnby-Borgstrom. 2003. The effects of pictures of emotional faces on tonic and phasic autonomic cardiac control in women and men. *Biological Psychology* 62:157–173.

Jung, C. 1961. *Memories, Dreams, Reflections.* Translated by A. Jaffe. New York: Vintage.

Juriloff, D. M., and M. J. Harris. 1991. Mapping the mouse craniofacial mutation first arch to chromosome two. *Journal of Heredity* 82:402–405.

Kaasinen, V., K. Nagren, J. Hietala, L. Farde, and J. O. Rinne. 2001. Sex differences in extrastriatal dopamine D(2)-like receptors in the human brain. *American Journal of Psychiatry* 158:308–311.

Kagan, J. 1971. *Change and Continuity in Infancy.* New York: Wiley.

———. 1994. *Galen's Prophecy.* New York: Basic Books.

Kagan, J., D. Arcus, N. Snidman, W. Yufeng, J. Hendler, and S. Green. 1994. Reactivity in infants: A cross national comparison. *Developmental Psychology* 30:342–345.

Kagan, J., E. Herbener, and J. Little. 1987. Temperamental features and differential threshold for butanol. Unpublished.

Kagan, J., R. B. Kearsley, and P. R. Zelazo. 1978. *Infancy: Its Place in Human Development.* Cambridge, MA: Harvard University Press.

Kagan, J., and H. A. Moss. 1962. *Birth to Maturity: A Study in Psychological Development*. New York: Wiley (reprinted by Yale University Press, 1982).

Kagan, J., and K. J. Saudino. 2001. Behavioral inhibition and related temperaments. In R. N. Emde and J. K. Hewitt, eds., *Infancy to Early Childhood*. New York: Oxford University Press, pp. 111–119.

Kagan, J., and N. Snidman. 1999. Early childhood predictors of adult anxiety disorder. *Biological Psychiatry* 46:1536–1541.

Kagan, J., N. Snidman, M. Zentner, and E. Peterson. 1999. Infant temperament and anxious symptoms in school-age children. *Development and Psychopathology* 11:209–224.

Kahn, I., Y. Yeshurun, P. Rotshtein, I. Freid, D. Ben-Bashat, and T. Hendler. 2002. The role of the amygdala in signaling perspective outcome of choice. *Neuron* 33:983–994.

Kajiwara, R., I. Takashima, Y. Mimura, M. P. Witter, and T. Iijima. 2003. Amygdala input promotes spread of excitatory neural activity from perirhinal cortex to the entorhinal-hippocampal circuit. *Journal of Neurophysiology* 89:2176–2184.

Kaplan, P. S., K. B. Fox, and E. R. Huckeby. 1992. Faces as reinforcers. *Developmental Psychobiology* 25:299–312.

Kapp, B. S., W. F. Supple, and P. J. Whalen. 1994. Effects of electrical stimulation of the amygdaloid central nucleus on neocortical arousal in the rabbit. *Behavioral Neuroscience* 108:81–93.

Kaviani, H., J. A. Gray, S. A. Checkley, V. Kumari, and G. D. Wilson. 1999. Modulation of the acoustic startle reflex by emotionally toned film-clips. *International Journal of Psychophysiology* 32:47–54.

Keightley, M. L., G. Winocur, S. J. Graham, H. S. Mayberg, S. J. Hevenor, and C. L. Grady. 2003. An fMRI study investigating cognitive modulation of brain regions associated with emotional processing of visual stimuli. *Neuropsychologia* 41:585–596.

Kellett, J., J. S. Marzillier, and C. Lambert. 1981. Social skill and somatotype. *British Journal of Medical Psychology* 54:149–155.

Keltner, D., and G. A. Bonnano. 1997. A study of laughter and dissociation. *Journal of Personality and Social Psychology* 73:687–702.

Kendler, K. S., J. M. Myers, and M. C. Neale. 2000. A multidimensional twin study of mental health in women. *American Journal of Psychiatry* 157:506–517.

Kerr, J. E., S. G. Beck, and R. J. Handa. 1996. Androgens selectively modulate C-fos messenger RNA induction in the rat hippocampus following novelty. *Neuroscience* 74:757–766.

Kim, M. S., J. J. Kim, and J. S. Kwon. 2001. Frontal P300 decrement and executive dysfunction in adolescents with conduct problems. *Child Psychiatry and Human Development* 32:93–106.

Kimble, M., D. Kaloupek, M. Kaufman, and P. Deldin. 2000. Stimulus novelty differentially affects attentional allocation in PTSD. *Biological Psychiatry* 47:880–890.

King, M. C., and A. G. Motulsky. 2002. Mapping human history. *Science* 298:2342–2343.

Kippin, T. E., S. W. Cain, and J. G. Pfaus. 2003. Estrous odors and sexually conditioned novel odors activate neural pathways in the male rat. *Neuroscience* 117(4):971–979.

Kistler, A., C. Mariauzouls, and K. von Berlepsch. 1998. Fingertip temperature as an indicator for sympathetic responses. *International Journal of Psychophysiology* 29:35–41.

Klepper, A., and H. Herbert. 1991. Distribution and origin of noradrenergic serotonergic fibers in the cochlear nucleus in the inferior colliculus of the rat. *Brain Research* 557:190–201.

Kline, J. P., G. C. Blackhart, K. M. Woodward, S. R. Williams, and G. Schwartz. 2000. Anterior electroencephalographic asymmetry changes in elderly women in response to a pleasant and unpleasant odor. *Biological Psychology* 52:241–250.

Kluver, H., and P. C. Bucy. 1939. Preliminary analysis of functions of the temporal lobe in monkeys. *Archives of Neurology and Psychiatry* 42:979–997.

Knight, R. T. 1996. Contribution of human hippocampal lesion to novelty detection. *Nature* 383:256–259.

Kochanska, G., K. C. Coy, T. L. Tjebkes, S. J. Husarek. 1998. Individual differences in emotionality in infancy. *Child Development* 69:375–390.

Korzan, W. J., T. R. Summers, and C. H. Summers. 2002. Manipulation of visual sympathetic sign stimulus modifies social status and plasma catecholamines. *General and Comparative Endocrinology* 128:153–161.

Kotani, S., S. Kawahara, and Y. Kirino. 2002. Classical eyeblink conditioning in decerebrate guinea pigs. *European Journal of Neuroscience* 15:1267–1270.

Koukounas, E., and R. Over. 2000. Changes in the magnitude of the eye blink startle reponse during habituation of sexual arousal. *Behaviour Research and Therapy* 38:573–584.

Kruska, D. 1988. Mammalian domestication and its effect on brain structure and behavior. In H. J. Jerison and I. Jerison, eds., *Intelligence and Evolutionary Biology,* NATO ASI Series, 617. Berlin: Springer-Verlag, pp. 211–250.

Kuhn, T. S. 2000. *The Road since Structure.* Chicago: University of Chicago Press.

Kumakiri, C., K. Kodama, E. Shimizu, N. Yamanouchi, S. Okada, S. Noda, H. Okamoto, T. Sato, and H. Shirasawa. 1999. Study of the association between the serotonin transporter gene regulating polymorphism and personality traits in a Japanese population. *Neuroscience Letters* 263:205–207.

Kwon, H. Y., S. J. Bultman, C. Loffler, W. J. Chen, P. J. Furdon, J. G. Powell, A. L. Usala, W. Wilkison, I. Hansmann, and R. P. Woychik. 1994. Molecular structure and chromosomal mapping of the human homolog of the agouti gene. *Proceedings of the National Academy of Sciences of the United States of America* 91:9760–9764.

Laakso, A., H. Vilkman, J. Bergman, M. Haaparanta, O. Solin, E. Syvalahti, R. K. Salokangas, and J. Hietala. 2002. Sex differences in striatal presynaptic dopamine synthesis capacity in healthy subjects. *Biological Psychiatry* 52:759–763.

La Bar, K. S., C. Gatenby, J. C. Gore, J. E. Le Doux, and E. A. Phelps. 1998. Human amygdala activation during conditioned fear acquisition and extinction. *Neuron* 29:937–945.

LaGasse, L., C. Gruber, and L. P. Lipsitt. 1989. The infantile expression of avidity in relation to later assessments. In J. S. Reznick, ed., *Perspectives on Behavioral Inhibition.* Chicago: University of Chicago Press, pp. 159–176.

La Greca, A. M., W. K. Silverman, and S. B. Wassastein. 1998. Children's predisaster functioning as a predictor of post-traumatic stress following Hurricane Andrew. *Journal of Consulting and Clinical Psychology* 66:883–892.

Lang, P. J., M. M. Bradley, and B. N. Cuthbert. 1992. A motivational analysis of emotion. *Psychological Review* 97:377–395.

Larsen, J. T., C. J. Norris, and J. T. Cacioppo. 2003. Effects of positive and negative affect on electromyographic activity over zygomaticus major and corrugator supercilli. *Psychophysiology* 40:776–785.

Larson, C. L., D. Ruffalo, J. Y. Nietert, and R. J. Davidson. 2000. Temporal stability of the emotion modulated startle response. *Psychophysiology* 37:92–101.

LeDoux, J. E. 1996. *The Emotional Brain.* New York: Simon and Schuster.

Lee, Y., D. E. Lopez, E. G. Meloni, and M. Davis. 1996. A primary acoustic startle circuit. *Journal of Neuroscience* 16:3775–3789.

Legrain, V., R. Bruyer, J. M. Guerit, and L. Plaghki. 2003. Nociceptive processing in the human brain of infrequent task-relevant and task-irrelevant noxious stimuli. *Pain* 103:237–248.

Leitner, D. S., and M. E. Cohen. 1985. Role of the inferior colliculus in the inhibition of the acoustic startle in the rat. *Physiology and Behavior* 34:65–70.

Lethbridge, R., J. G. Simmons, and N. B. Allen. 2002. All things pleasant are not equal: Startle reflex modification while processing social and physical threat. *Psychophysiology* 39, S51.

Levine, R. A. 2003. The kindness of strangers. *American Scientist* 91:226–233.

Lewis, M., D. S. Ramsay, and K. Kawakami. 1993. Differences between Japanese infants and Caucasian–American infants in behavioral and cortisol response to inoculation. *Child Development* 64:1722–1731.

Lewontin, R. 1995. *Human Diversity.* New York: Scientific American.

Li, L., P. M. Prieber, and J. S. Yeomans. 1998. Prepulse inhibition of acoustic or trigeminal startle of rats by unilateral electrical stimulation of the inferior colliculus. *Behavioral Neuroscience* 112:1187–1198.

Lin, K. M., R. E. Poland, and I. M. Lesser. 1986. Ethnicity and psychopharmacology. *Culture Medicine and Psychiatry* 10:151–165.

Lipp, O. V., and D. A. T. Siddle. 1999. Startle modification during orienting and Pavlovian conditioning. In M. E. Dawson, A. M. Schell, and A. H. Bohmelt, eds., *Startle Modification.* New York: Cambridge University Press, pp. 300–313.

Lipp, O. V., D. A. T. Siddle, and P. J. Dall. 1997. The effect of emotional and attentional processes to blink startle modulation and on electrodermal processes. *Psychophysiology* 34:340–347.

———. 2000. The effect of warning stimulus modality on blink startle modification in reaction time tasks. *Psychophysiology* 37:55–64.

Lloyd, R. L., and A. S. Kling. 1991. Delta activity from the amygdala in squirrel monkeys (Sainiri sciureus). *Behavioral Neuroscience* 105:223–229.

Loewy, A. D. 1990a. Anatomy of the autonomic nervous system. In A. D. Loewy and K. M. Spyer, eds., *Central Regulation of Autonomic Functions.* New York: Oxford University Press, pp. 3–16.

———. 1990b. Central autonomic patterns. In A. D. Loewy and K. M. Spyer, eds., *Central Regulation of Autonomic Functions.* New York: Oxford University Press, pp. 88–103.

Lopes da Silva, F. H., T. H. van Lierop, C. Schrijer, and W. Strom van Leeuwen. 1973. Organization of thalamic and cortical alpha rhythm spectra and coherence. *Electroencephalography and Clinical Neurophysiology* 35:627–639.

Lu, D., D. Willard, I. R. Patel, S. Kadwell, L. Overton, T. Kost, M. Luther, W. Chen, R. P. Woychik, and W. O. Wilkison. 1994. Agouti protein is an antagonist of the melanocyte-stimulating-hormone receptor. *Nature* 371:799–802.

Lupie, S. J., S. King, M. J. Meaney, and B. S. McEwen. 2001. Can poverty get under your skin? *Development and Psychopathology* 13:653–676.

Lynn, R., and S. L. Hampson. 1977. Fluctuations in natural levels of neuroticism and extraversion 1935–1970. *British Journal of Social and Clinical Psychology* 16:131–137.

Lyons, P. M., H. Afarion, A. F. Schatzberg, A. Sawyer-Glover, and M. E. Moseley. 2002. Experience dependent asymmetric variation in monkeys. (Saimiri sciurencis). *Behavioral Brain Research* 136:51–59.

Macedo, C. E., V. M. Castilho, C. de Souza, M. A. Silva, and M. L. Brandao. 2002. Dual 5-HT mechanisms in basolateral and central nuclei of the amygdala in the regulation of the defensive behavior induced by electrical stimulation of the inferior colliculus. *Brain Research Bulletin* 59:189–195.

Magnusson, D. 2000. The individual as the organizing principle in psychological inquiry. In L. R. Bergman, R. B. Cairns, L. G. Nilsson, L. Nystedt, eds., *Developmental Science and the Holistic Psychological Profile.* Mahwah, NJ: Erlbaum, pp. 33–48.

Maisonnette, S. S., M. C. Kawasaki, N. C. Coimbra, and M. L. Brandao. 1996. Effects of lesions of the amygdaloid nuclei and substantia nigra on aversive response induced by electrical stimulation of the inferior colliculus. *Brain Research Bulletin* 40:93–98.

Makino, S., T. Shibasaki, N. Yamauchi, and T. Nishioka. 1999. Psychological stress increased corticotropin releasing hormone mRNA alpha content in the central nucleus of the amygdala but not in the hypothalamic paraventricular nucleus in the rat. *Brain Research* 250:136–143.

Manke, B., K. J. Saudino, and J. D. Grant. 2001. Extremes analyses of observed temperament dimensions. In R. N. Emde and J. P. Hewitt, eds., *Infancy to Early Childhood*. New York: Oxford University Press, pp. 52–72.

Maren, S. 2001. Neurobiology of Pavlovian fear conditioning. *Annual Review of Neuroscience* 24:897–931.

Marks, I. M. 1987. *Fears, Phobias, and Rituals*. New York: Oxford University Press.

Marsh, R. A., Z. M. Fuzessery, C. D. Grose, and J. J. Wenstrup. 2002. Projection to the inferior colliculus from the basolateral nucleus of the amygdala. *Journal of Neuroscience* 22:10449–10460.

Marshall, P. J., and J. Stevenson-Hinde. 1998. Behavioral inhibition, heart period, and respiratory sinus arrhythmia in young children. *Developmental Psychobiology* 33:283–292.

———. 2001. Behavioral inhibition, heart period, and respiratory sinus arrhythmia in young children. University of Cambridge, unpublished.

Martin, P. 2003. The epidemiology of anxiety disorders. *Dialogues in Clinical Neuroscience* 5:281–298.

Marui, T., O. Hashimoto, E. Nanba, C. Kato, M. Tochigi, T. Umekage, N. Kato, and T. Sasaki. 2004. Gastrin-releasing peptide receptor (GRPR) locus in Japanese subjects with autism. *Brain and Development* 26:5–7.

McDougall, W. 1929. The chemical theory of temperament applied to introversion and extraversion. *Journal of Abnormal and Social Psychology* 24:293–309.

McGaugh, J. L., and L. Cahill. 2003. Emotion and memory. In R. J. Davidson, K. R. Scherer, and H. H. Goldsmith, eds., *Handbook of Affective Science*. New York: Oxford University Press, pp. 93–116.

McManis, M. H., M. M. Bradley, W. K. Berg, B. N. Cuthbert, and P. J. Lang. 2001. Emotional reactions in children. *Psychophysiology* 38:222–231.

McNally, C. P., and H. Akil. 2002. Role of corticotropin-releasing hormone in the amygdala and bed nucleus of the stria terminalis in the behavioral, pain-modulating, and endocrine consequences of opiate withdrawal. *Neuroscience* 112:605–613.

Means, M. H. 1936. Fears of one thousand college women. *Journal of Abnormal and Social Psychology* 31:291–311.

Meller, S. T., and B. J. Dennis. 1991. Efferent projections of the periaqueductal gray in the rabbit. *Neuroscience* 40:191–216.

Meloni, D. G., and M. Davis. 1999. Muscimol in the deep layer of the superior colliculs/mesencephalic reticular formation block expression but not acquisition of fear-potentiated startle in rats. *Behavioral Neuroscience* 113:1152–1160.

Merikangas, K. R., and N. Risch. 2003. Will the genomics revolution revolutionalize psychiatry? *American Journal of Psychiatry* 160:625–635.

Meyer, G. J., S. E. Finn, L. D. Eyde, G. G. Kay, K. L. Moreland, R. R. Dies, E. J. Eisman, T. W. Kubiszyn, and G. M. Reed. 2001. Psychological testing and psychological assessment. *American Psychologist* 56:128–165.

Miller, M. W., and J. L. Greif. 2002. Is startle exaggerated in post-traumatic stress disorder? Personal Communication.

Miller, M. W., C. J. Patrick, and G. K. Levenson. 2002. Affective imagery and the startle response. *Psychophysiology* 39:515–529.

Mills, A. D., and J. M. Faure. 1991. Diversion selection for duration of chronic immobility and social reinstatement behavior in Japanese quail (Coturnix Japonica) chicks. *Journal of Comparative Psychology* 105:25–38.

Miltner, W., M. Matjak, C. Braun, H. Diekmann, and S. Brody. 1994. Emotional qualities of odors and their influence on the startle reflex in humans. *Psychophysiology* 31:107–110.

Miranda, M. I., L. Ramirez-Lugo, and F. Bermudez-Rattoni. 2000. Cortical cholinergic activity is related to the novelty of the stimulus. *Brain Research* 882:230–235.

Mischel, W. 2004. Toward an integrative science of the person. In S. T. Fiske, D. L. Schacter, and C. Zahn-Waxler, eds., *Annual Review of Psychology* 55:1–22.

Miwa, H., C. Nohara, M. Hotta, and Y. Shimo. 1998. Somatosensory evoked blink responses. *Brain* 121:281–291.

Miyawaki, T., A. K. Goodchild, and P. M. Pilowsky. 2002. Activation of mu-opioid receptors in rat ventrolateral medulla selectively blocks baroreceptor reflexes while activation of delta opioid receptors blocks somato-symapathetic reflexes. *Neuroscience* 109:133–144.

Moelle, M., L. Marshall, H. L. Fehm, and J. Born. 2002. EEG theta synchronization conjoined with alpha desynchronization indicate intentional encoding. *European Journal of Neuroscience* 15:923–928.

Moller, A. R., P. J. Jannetta, and H. D. Jho. 1994. Click-evoked responses from the cochlear nucleus. *Electroencephalography and Clinical Neurophysiology* 92:215–224.

Morley-Fletcher, S., P. Palanza, D. Parolaro, D. Vigaero, and G. Laviola. 2003. Intra-uterine position has long term influences on brain μ-opioid receptor densities and behavior in mice. *Psychoneuroendocrinology* 28:386–400.

Mozley, L. H., R. C. Gur, P. D. Mozley, and R. E. Gur. 2001. Striatal dopamine transporters and cognitive functioning in healthy men and women. *American Journal of Psychiatry* 158:1492–1499.

Mulken, S., S. M. Bogels, P. J. de Jong, and J. Louwers. 2001. Fear of blushing. *Journal of Anxiety Disorders* 15:413–432.

Munoz-Blanco, J., and A. P. Castillo. 1987. Changes in neurotransmitter amino acids content in several CNS areas from aggressive and non-agressive bull strains. *Physiology and Behavior* 39:453–457.

Muris, P., H. Merckelbach, B. Mayer, and E. Prins. 2000. How serious are common childhood fears? *Behavioral Research and Therapy* 38:217–228.

Naatanen, R., and I. Winkler. 1999. The concept of auditory stimulus presentation in cognitive neuroscience. *Psychological Bulletin* 125:826–859.

Nagae, S., and M. Moscovitch. 2002. Cerebral hemisphere differences in memory of emotional and nonemotional words in normal individuals. *Neuropsychologia* 40:1601–1607.

Nelson, C. A., and C. S. Monk. 2001. The use of event-related potentials in the study of cognitive development. In C. A. Nelson and M. Luciana, eds., *Handbook of Developmental Cognitive Neuroscience.* Cambridge, MA: MIT Press, pp. 125–136.

Nelson, C. A., K. M. Thomas, M. de Haan, and S. S. Wewerka. 1998. Delayed recognition memory in infants and adults as revealed by event-related potentials. *International Journal of Psychophysiology* 29:145–165.

Nelson, E. E., S. E. Shelton, and N. H. Kalin. 2003. Individual differences in the responses of naive rhesus monkeys to snakes. *Emotion* 3:3–11.

Newell, R., and I. Marks. 2000. The phobic nature of social difficulty in facially disfigured people. *British Journal of Psychiatry* 176:177–181.

Niedenthal, P. M., J. B. Halberstadt, and A. H. Innes-Kerr. 1999. Emotional response categorization. *Psychological Review* 106:337–361.

Nijhout, H. F. 2003. The importance of context in genetics. *American Scientist* 91:416–423.

Nisbett, R. E., K. Peng, I. Choi, and A. Norenzayan. 2001. Culture and systems of thought. *Psychological Review* 108:291–310.

Nitschke, J. B., C. L. Larson, M. J. Smoller, S. D. Navin, A. J. C. Pederson, D. Ruffalo, K. L. Mackiewicz, S. M. Gray, E. Victor, and R. J. Davidson. 2002. Startle potentiation in aversive anticipation. *Psychophysiology* 39:254–258.

Ochsner, K. N. 2000. Are affective events richly recollected or singly familiar? *Journal of Experimental Psychology: General* 129:242–261.

O'Doherty, J., J. Winston, H. Critchley, D. Perrett, D. M. Burt, and R. J. Dolan. 2003. Beauty in a smile. *Neuropsychologia* 41:147–155.

Ogembo, J. M. 2001. Cultural narratives, violence, and mother-son loyalty. *Ethos* 29:3–29.

Ohman, A., and S. Mineka. 2001. Fears, phobias, and preparedness. *Psychological Review* 108:483–522.

———. 2003. The malicious serpent. *Current Directions in Psychological Science* 12:5–9.

Olivares, E. I., J. Iglesias, and S. Rodriguez-Holguin. 2003. Long-latency ERPs and recognition of facial identity. *Journal of Cognitive Neuroscience* 15(1):136–151.

Onu, S., and H. Nishijo. 2000. Neruophysiological bases of emotion in primates. In M. Gazzaniga, ed., *New Cognitive Neuroscience*. 2nd ed. Cambridge, MA: MIT Press, pp. 1099–1114.

Ornitz, E. M., J. G. Jehricke, A. T. Russell, R. Pynoos, and P. Siddarth. 2001. Modulation of startle and the startle elicited P300 by the conditions of the cued continuous performance task in school-age boys. *Clinical Neurophysiology* 112:2209–2223.

Orr, S. P., Z. Solomon, T. Peri, R. K. Pitman, and A. Y. Shalev. 1997. Physiologic response to loud tones in Israeli veterans of the 1973 Yom Kippur War. *Biological Psychiatry* 41:319–326.

Ozer, E. J., S. R. Best, T. C. Lipsey, and D. S. Weiss. 2003. Predictors of posttraumatic stress disorders and symptoms in adults. *Psychological Bulletin* 129:52–73.

Page, A. C. 2003. The role of disgust in faintness elicited by blood and injection stimuli. *Journal of Anxiety Disorders* 17:45–58.

Palomba, D., M. Sarlo, A. Angrilli, A. Mini, and L. Stegagno. 2000. Cardiac responses associated with affective processing of unpleasant film stimuli. *International Journal of Psychophysiology* 36:45–57.

Pandossio, J. E., V. A. Molina, and M. L. Brandao. 2000. Prior electrical stimulation of the inferior colliculus sensitizes rats to the stress of the elevated plus-maze test. *Behavioral Brain Research* 109:19–25.

Papousek, I., and G. Schulter. 1998. Different temporal stability and partial independence of EEG asymmetries for different locations. *International Journal of Neuroscience* 93:87–100.

Park, L., and D. Hinton. 2002. Dizziness and panic in China. *Culture, Medicine, and Psychiatry* 26:225–257.

Pavlov, I. 1927. *Conditioned Reflexes*. London: Oxford University Press.

Petren, S., M. A. Munaw, K. Keune, and D. L. Philion. 2002. Personality dimensions of emotional inhibition and extraversion in predicting the emotional modulation of startle. *Psychophysiology* 39: Supplement 1, p. S65.

Petrovich, G. D., N. S. Canteras, and L. W. Swanson. 2001. Combinatorial amygdalar inputs to hippocampal and hypothalamic behavioral systems. *Brain Research Reviews* 38:247–289.

Pfeifer, M., H. H. Goldsmith, R. J. Davidson, and M. Rickman. 2002. Continuity and change in inhibited and uninhibited children. *Child Development* 73:1474–1485.

Piefke, M., P. H. Weiss, K. Zilles, H. J. Markowitsch, and G. R. Fink. 2003. Differential remoteness and emotional tone modulate the neural correlates of autobiographical memory. *Brain* 126:650–668.

Pitkanen, A. 2000. Connectivity of the rat amygdaloid complex. In J. P. Aggleton, ed., *The Amygdala: A Functional Analysis*. 2nd ed. London: Oxford University Press, pp. 31–116.

Pizzagalli, D., T. Koenig, M. Regard, and D. Lehmann. 1998. Faces and emotions. *Neuropsychologia* 36:323–332.

Pizzagalli, D. A., J. B. Nitschke, T. R. Oakes, A. M. Hendrick, K. A. Horras, C. L. Larson, H. C. Abercrombie, S. M. Schaefer, J. V. Koger, R. M. Benca, R. D. Pascual-Marqui, and R. J. Davidson. 2002. Brain electrical tomography in depression. *Biological Psychiatry* 52:73–85.

Plotkin, H. L. 1988. Behavior and evolution. In H. L. Plotkin, ed., *The Role of Behavioral Evolution.* Cambridge, MA: MIT Press, pp. 1–18.

Porges, S. W., J. A. Doussard-Roosevelt, A. L. Portales, and P. E. Suess. 1994. Cardiac vagal tone. *Developmental Psychobiology* 27(5):289–300.

Price, J. L. 1999. Prefrontal cortical networks related to visceral function and mood. *Annals of the New York Academy of Sciences* 877:383–396.

Price, J. L., and D. G. Amaral. 1981. An autoradiographic study of the projections of the central nucleus of the amygdala. *Journal of Neuroscience* 1:1242–1259.

Pujol, J., A. Lopez, J. Deus, N. Cardoner, J. Vallejo, A. Capdevila, and T. Paus. 2002. Anatomical variability of the anterior cingulate gyrus and basic dimensions of human personality. *Neuroimage* 15:847–855.

Pullias, E. V. 1937. Masturbation as a mental hygiene problem. *Journal of Abnormal and Social Psychology* 32:216–222.

Pyles, M. K., H. R. Stolz, and J. W. MacFarlane. 1935. The accuracy of mothers' reports on birth and developmental data. *Child Development* 6:165–176.

Pynoos, R. S., C. Frederick, K. Nader, W. Arroyo, A. Steinberg, S. Feth, F. Nunez, and L. Fairbanks. 1987. Life threat and post-traumatic stress disorder in school-age children. *Archives of General Psychiatry* 44:1057–1063.

Ramnani, N., I. Toni, O. Josephs, J. Ashburner, and R. E. Passingham. 2000. Learning and expectation related changes in the human brain during motor learning. *Journal of Neurophysiology* 84:3026–3035.

Ray, W. J., and H. W. Cole. 1985. EEG alpha activity reflects attentional demands and beta activity reflects emotional and cognitive processes. *Science* 228:750–752.

Rebec, G. V., J. R. Christianson, C. Guerra, and M. T. Bardo. 1997a. Regional and temporal differences in real time dopamine efflux in the nucleus accumbens during food choice novelty. *Brain Research* 776:61–67.

Rebec, G. V., C. P. Grabner, M. Johnson, R. C. Pierce, and M. T. Bardo. 1997b. Transient increases in catecholaminergic activity in medial prefrontal cortex and nucleus accumbens shell. *Neuroscience* 76:707–714.

Ren, Z. G., P. D. Porzgen, Y. H. Youn, and M. Sieber-Blum. 2003. Ubiquitous embryonic expression of the norepinephrine transporter. *Developmental Neuroscience* 25:1–13.

Reppucci, C. 1968. Hereditary influences on the distribution of attention in infancy. PhD diss., Harvard University.

Reznick, J. S., J. L. Gibbons, M. O. Johnson, and P. M. McDonough. 1989. Behavioral inhibition in a normative sample. In J. S. Reznick, ed., *Perspectives on Behavioral Inhibition*. Chicago: University of Chicago Press, pp. 25–49.

Reznick, J. S., J. Kagan, N. Snidman, M. Gersten, K. Baak, and A. Rosenberg. 1986. Inhibited and uninhibited children. *Child Development* 57:660–680.

Rich, G. J. 1928. A biochemical approach to the study of personality. *Journal of Abnormal and Social Psychology* 23:158–175.

Richards, A., C. L. French, A. J. Calder, B. Webb, R. Fox, and A. W. Young. 2002. Anxiety-related bias in the classification of the centrally ambiguous facial expressions. *Emotion* 2:273–287.

Richards, J. E. 2000. Development of multimodal attention in young infants. *Psychophysiology* 37:65–75.

Richardson, R., and P. McNally. 2003. Effects of an odor paired with illness on startle, freezing, and analgesia in rats. *Physiology and Behavior* 78:213–219.

Rickman, M. D. 1998. Behavioral inhibition, emotional vulnerability and brain asymmetry. *Dissertation Abstracts International: Section B; Sciences and Engineering* 59 (2-B), (August).

Rimm-Kaufman, S. E., D. M. Early, M. J. Cox, G. Saluja, R. C. Pianta, R. H. Bradley, and C. Payne. 2002. Early behavior attributes and teachers' sensitivity as predictors of competent behavior in the kindergarten classroom. *Applied Developmental Psychology* 23:451–470.

Robinson, J. L., J. Kagan, J. S. Reznick, and R. Corley. 1992. The heritability of inhibited and uninhibited behavior. *Developmental Psychology* 28:1030–1037.

Roitman, M. F., G. Van Dijk, T. E. Thiele, and I. L. Bernstein. 2001. Dopamine mediation of the feeding response to violations of spatial and temporal expectations. *Behavioral Brain Research* 122:193–199.

Rolls, E. T., J. O'Doherty, M. L. Kringelbach, S. Francis, R. Bowtell, and

F. McGlone. 2003. Representations of pleasant and painful touch in the human orbitofrontal and cingulate cortices. *Cerebral Cortex* 13:308–317.

Roozendaal, B., and A. R. Cools. 1994. Influence of the noradrenergic state of the nucleus accumbens in basolateral amygdala mediated changes in neophobia of rats. *Behavioral Neuroscience* 108:1107–1118.

Roschmann, R., and W. Wittling. 1992. Topographic brain mapping of emotion related hemisphere asymmetry. *International Journal of Neuroscience* 63:5–16.

Rosenbaum, J. F., J. Biederman, D. R. Hirschfeld-Becker, J. Kagan, N. Snidman, D. Friedman, A. Nineberg, D. J. Gallery, and S. V. Faraone. 2000. A control study of behavioral inhibition in children of parents of panic disorder and depression. *American Journal of Psychiatry* 157:2002–2010.

Rosenberg, A., and J. Kagan. 1987. Iris pigmentation and behavioral inhibition. *Developmental Psychobiology* 20:377–392.

Rosenblum, L. A., E. L. Smith, M. Altemus, B. A. Scharf, M. J. Owens, C. B. Nemeroff, J. M. Gorman, and J. D. Coplan. 2002. Differing concentrations of corticotropin-releasing factor and oxytocin in the cerebrospinal fluid of bonnet and pigtail macques. *Psychoneuroendocrinology* 27:651–660.

Rosenkranz, J. A., and A. A. Grace. 2002. Dopamine-mediated modulation of odor evoked amygdala potentials during Pavlovian conditioning. *Nature* 417:282–287.

Røskaft, E., T. Bjerke, B. Kaltenborn, J. D. C. Linnell, and R. Andersen. 2003. Patterns of self-reported fear towards large carnivores among the Norwegian public. *Evolution and Human Behavior* 24:184–198.

Rothbart, M. K. 1989. Temperament in childhood. In J. A. Kohnstamm, J. E. Bates, and M. K. Rothbart, eds., *Temperament in Childhood.* New York: Wiley, pp. 59–76.

Rothbart, M. K., S. A. Ahadi, K. L. Hershey, and P. Fisher. 2001. Investigations of temperament at three to seven years. *Child Development* 72:1394–1408.

Rothbart, M. K., L. K. Ellis, M. R. Rueda, and M. I. Posner. 2003. Developing mechanisms of temperamental effortful control. *Journal of Personality* 71:1113–1143.

Rubin, K. H., and L. Both. 1989. Iris pigmentation and sociability in childhood. *Developmental Psychobiology* 22:1–9.

Rubin, K. H., K. B. Burgess, and D. D. Hastings. 2002. Stability and social behavioral consequences of toddlers' inhibited temperament and parenting behaviors. *Child Development* 73:483–495.

Rubin, K. H., P. D. Hastings, S. L. Stewart, H. A. Henderson, and X. Chen. 1997. The consistency and concomitants of inhibition. *Child Development* 68:467–483.

Saavedra, L. M., and W. K. Silverman. 2002. Case study: disgust and a specific phobia of buttons. *Journal of the American Academy of Child and Adolescent Psychiatry* 41:1376–1379.

Schaul, N. 1998. The fundamental neural mechanisms of electroencephalography. *Electroencephalography and Clinical Neurophysiology* 106:101–107.

Schelde, J. T. 1998. Major depression: Behavioral markers of depression and recovery. *Journal of Nervous and Mental Disease* 186:133–140.

Scherer, K. R. 1997. Profiles of emotional antecedent appraisal. *Cognition and Emotion* 11:113–150.

Schmid, N. B., J. P. Forsyth, H. T. Santiago, and J. H. Trakowski. 2002. Classification of panic attack subtypes in patients and normal controls in response to biological challenge. *Journal of Anxiety Disorders* 16:625–638.

Schmidt, L. A. 1999. Frontal brain electrical activity in shyness and sociability. *Psychological Science* 10:316–320.

Schmidt, L. A., K. A. Cote, D. L. Santesso, and C. E. Milner. 2003. Frontal electroencephalogram alpha asymmetry during sleep: stability and its relation to affective style. *Emotion* 3:401–407.

Schmidt, L. A., N. A. Fox, J. Schulkin, and P. W. Gold. 1999. Behavioral and psychophysiological correlates of self-presentation in temperamentally shy children. *Developmental Psychobiology* 35:119–135.

Schmidt, L. A., L. J. Trainor, and D. L. Santesso. 2003. Development of frontal electroencephalogram (EEG) and heartrate (ECG) responses to affective musical stimuli during the first twelve months of postnatal life. *Brain and Cognition* 52:27–32.

Schmidt, S. R. 2002. Outstanding memories. *Journal of Experimental Psychology: Learning Memory and Cognition* 28:353–361.

Schmidtke, K., and J. A. Buttner-Ennever. 1992. Nervous control of eyelid function. *Brain* 115:227–247.

Schnitzler, A., J. Volkmann, P. Enck, T. Frieling, O. W. Witte, and H. J. Freund. 1999. Different cortical organization of visceral and somatic sensation in humans. *European Journal of Neuroscience* 11:305–315.

Schultz, W. 2002. Getting formal with dopamine and reward. *Neuron* 36:241–263.

Schwartz, C. E., N. Snidman, and J. Kagan. 1999. Adolescent social anxiety and outcome of inhibited temperament in childhood. *Journal of the American Academy of Child and Adolescent Psychiatry* 38:1008–1015.

Schwartz, C. E., C. I. Wright, L. M. Shin, J. Kagan, and S. L. Rauch. 2003a. Inhibited and uninhibited infants "grown up": Adult amygdalar response to novelty. *Science* 300:1952–1953.

Schwartz, C. E., C. I. Wright, L. M. Shin, J. Kagan, P. J. Whalen, K. G. McMullin, and S. L. Rauch. 2003b. Differential amygdalar response to novel versus newly familiar neutral faces. *Biological Psychiatry* 53:854–862.

Schwarz, N. 1999. Self-reports. *American Psychologist* 54:93–105.

Scott, J. P., and J. Fuller. 1965. *Genetics and the Social Behavior of the Dog*. Chicago: University of Chicago Press.

Seidenbecher, T., T. R. Laxmi, O. Stork, and H. C. Pape. 2003. Amygdalar and hippocampal theta rhythm synchronization during fear memory retrieval. *Science* 301:846–850.

Seifer, R., A. J. Sameroff, and E. Krafchuk. 1994. Infant temperament measured by multiple observations and mother report. *Child Development* 65:1478–1490.

Seyfarth, R. M., and D. L. Cheney. 2003. Signalers and receivers in animal communication. In S. T. Fiske, D. L. Schacter, and C. Zahn-Waxler, eds., *Annual Review of Psychology* 54:145–173.

Shamni, P., and D. T. Stuss. 1999. Humor appreciation. *Brain* 122:657–666.

Shaw, D. S., M. Gillion, E. M. Ingoldsby, and P. S. Nagio. 2003. Trajectories leading to school-age conduct problems. *Developmental Psychology* 39:189–200.

Shumyatsky, G. P., E. Tsvetkov, G. Malleret, S. Vronskaya, M. Hatton, L. Hampton, J. F. Battey, C. Dulac, E. R. Kandel, and V. Y. Bolshakov. 2002. Identification of a signaling network in lateral nucleus of amygdala important for inhibiting memory specifically related to learned fear. *Cell* 11:905–918.

Silva, J. R., D. A. Pizzagalli, C. L. Lawson, D. L. Jackson, and R. J. Davidson. 2002. Frontal brain asymmetry in restricted eaters. *Journal of Abnormal Psychology* 111:676–681.

Skolnick, A. J., and R. J. Davidson. 2002. Affective modulation of eye blink startle with reward and threat. *Psychophysiology* 39:835–850.

Smoller, J. W., J. F. Rosenbaum, J. Biederman, J. Kennedy, D. Dai, S. Racette, N. Laird, J. Kagan, N. Snidman, D. Hirshfeld-Becker, M. T. Tsuang, P. B. Sklar, and S. A. Slaugenhaupt. 2003. Association of a genetic marker at the corticotropin-releasing hormone locus with behavioral inhibition. *Biological Psychiatry* 54(12):1376–1381.

Snidman, N. 1989. Behavioral inhibition and sympathetic influence on the cardiovascular system. In J. Reznick, ed., *Perspectives on Behavioral Inhibition*. Chicago: University of Chicago Press, pp. 51–70.

Snidman, N., J. Kagan, L. Riordan, and D. Shannon. 1995. Cardiac function and behavioral reactivity during infancy. *Psychophysiology* 32:199–207.

Song, T. M., L. Perusse, R. A. Molina, and C. Bouchard. 1994. Twin resemblance in somatotype in comparison with other twin studies. *Human Biology* 66:453–464.

Sorensen, D., M. McManis, and J. Kagan. Unpublished. Determinants of startle.

Sorokin, P. 1924. *Leaves from a Russian Diary*. New York: E. P. Dutton, p. 189.

Stams, G. J. M., F. Juffer, and A. H. van IJzendoorn. 2002. Maternal sensitivity, infant attachment, and temperament in early childhood predict adjustment in middle childhood. *Developmental Psychology* 38:806–821.

Stelmack, R. M. 1990. Biological bases of extraversion. *Journal of Personality* 58:293–311.

Stelmack, R. M., V. Knott, and C. M. Beauchamp. 2003. Intelligence and neural transmission time. *Personality and Individual Differences* 34:97–107

Stevenson-Hinde, J., and P. J. Marshall. 1999. Behavioral inhibition, heart period, and respiratory sinus arrhythmia. *Child Development* 70:805–816.

Stifter, C. A., N. A. Fox, and S. W. Porges. 1989. Facial expressivity and vagal tone in 5- and 10-month old infants. *Infant Behavior and Development* 12:127–137.

Stifter, C. A., and A. Jain. 1996. Psychophysiological correlates of infant temperament. *Developmental Psychobiology* 29:379–391.

Stroud, L. R., P. Salovey, and E. S. Epel. 2002. Sex differences in stress responses. *Biological Psychiatry* 52:318–327.

Stroud, L. R., M. Tanofsky-Kraff, D. E. Wilfley, and P. Salovey. 2000. The Yale Interpersonal Stressor (YIPS). *Annals of Behavioral Medicine* 22:204–213.

Sullivan, R. M., and A. Gratton. 2002. Behavioral affects of excitotoxic lesions of ventral medial prefrontal cortex in the rat are hemisphere dependent. *Brain Research* 927:69–79.

Sulloway, F. J. 1996. *Born to Rebel*. New York: Pantheon.

Sundet, J. M., I. Skre, J. J. Okkenhaug, and K. Tambs. 2003. Genetic and environmental causes of the interrelationships between self-reported fears. *Scandinavian Journal of Psychology* 44:97–106.

Sutton, S. K., and R. J. Davidson. 2000. Prefrontal brain electrical asymmetry predicts the evaluation of affective stimuli. *Neuropsychologia* 38:1723–1733.

Sutton, S. K., R. J. Davidson, B. Donzella, W. Irwin, and D. A. Dottl. 1997. Manipulating affect state using extended picture presentations. *Psychophysiology* 34:217–226.

Swickert, R. J., and K. Gilliland. 1998. Relationship between the brainstem auditory evoked response and extraversion, introversion, impulsivity, and sociability. *Journal of Research in Personality* 3:373–380.

Tang, A. C., and B. Zou. 2002. Neonatal exposure to novelty enhances long term potentiation in CA1 of the rat hippocampus. *Hippocampus* 12:398–404.

Tang, J., C. T. Wotjak, S. Wagner, G. Williams, M. Schachner, and A. Dityatev. 2001. Potentiated amygdaloid auditory-evoked potentials and freezing behavior after fear conditioning in mice. *Brain Research* 919:232–241.

Tavernov, S. J., K. A. J. Abduljawad, R. W. Langley, and C. M. Bradshaw. 2000. Effects of pentagastrin and the cold pressor test on the acoustic startle response and pupillary function in man. *Journal of Psychopharmacology* 14:387–394.

Temeles, E. J., and W. J. Kress. 2003. Adaptation in a plant-hummingbird association. *Science* 300:630–633.

Terr, L. C. 1979. Children of Chowchilla. *Psychoanalytic Study of the Child* 34:547–627.

Thomas, A., and S. Chess. 1963. *Behavioral Individuality in Early Childhood*. New York: New York University Press.

Thomas, A., S. Chess, and H. G. Birch. 1969. *Temperament and Behavioral Disorders in Children*. New York: New York University Press.

Thorndike, E. L. 1907. *The Elements of Psychology*. 2nd ed. New York: Seiler.

Tillfors, M., T. Furmark, and I. Marteindottir, and M. Fredrikson. 2002. Cerebral blood flow during anticipation of public speaking and social phobia. *Biological Psychiatry* 52:1113–1119.

Tomarken, A. J., R. J. Davidson, and J. B. Henriques. 1990. Resting frontal brain asymmetry predicts affective responses to films. *Journal of Personality and Social Psychology* 59:791–801.

Trayn, Y., A. Craig, and T. McIsaac. 2001. Extroversion-introversion and 8–13 Hz waves in the frontal cortical regions. *Personality and Individual Differences* 30:205–215.

Trut, L. N. 1999. Early canid domestication. *American Scientist* 87:160–169.

Tseng, W., K. Mo, J. Hsu, and L. Li. 1988. A socio-cultural study of Koro epidemics in Guangdong, China. *American Journal of Psychiatry* 145:1538–1543.

Turkheimer, E., A. Haley, M. Waldron, B. D'Onofrio, and I. I. Gottsman. 2003. Socioeconomic status modifies heritability of IQ in young children. *Psychological Science* 14:623–628.

Ueda, K., Y. Okamoto, G. Okada, H. Yamashita, T. Hori, and S.Yamawaki. 2003. Brain activity during expectancy of emotional stimuli. *Neuroreport* 14:51–55.

Valentine, C. W. 1930. The innate basis of fear. *Journal of Genetic Psychology* 37:394–420.

Valverde, P., E. Healy, I. Jackson, J. L. Rees, and A. J. Thody. 1995. Variants of the melanocyte-stimulating hormone receptor gene are associated with red hair and fair skin in humans. *Nature Genetics* 11:328–330.

Van Rollen, R., and S. J. Thorpe. 2001. The time course of visual processing. *Journal of Cognitive Neuroscience* 13:454–461.

Veit, R., H. Flor, M. Erb, C. Hermann, M. Lotze, W. Grodd, and N. Birbaumer. 2002. Brain circuits involved in emotional learning in antisocial behavior and social phobia in humans. *Neuroscience Letters* 328:233–236.

Vinogradova, O. S. 2001. Hippocampus as comparator. *Hippocampus* 11:578–598.

Vitalis, T., C. Alvarez, K. Chen, J. C. Shih, P. Gaspar, and O. Cases. 2003. Developmental expression pattern of monoamine oxidase in sensory organs and neural crest derivatives. *Journal of Comparative Neurology* 464:392–403.

Vrana, S. R., E. L. Spence, and P. J. Lang. 1988. The startle probe response. *Journal of Abnormal Psychology* 97:487–491.

Wacker, J., M. Heldmann, and G. Stemmler. 2003. Separating emotion and motivational direction in fear and anger. *Emotion* 3:167–193.

Walker, D. L., and M. Davis. 1997. Involvement of the dorsal periaqueductal gray in loss of fear potentiated startle accompanying high foot shock training. *Behavioral Neuroscience* 11:692–702.

Walker, D. L., D. J. Toufexis, and M. Davis. 2003. Role of the bed nucleus of the stria terminalis vs. the amygdala in fear, stress, and anxiety. *European Journal of Pharmacology* 463:199–216.

Wallace, R. J., and J. B. Rosen. 2001. Neurotoxic lesions of the lateral nucleus of the amygdala decrease conditioned fear but not unconditioned fear of a predator odor. *Journal of Neuroscience* 21:3619–3627.

Wallis, R. S. 1954. The overt fears of Dakota Indian children. *Child Development* 25:185–192.

Wang, H., and M. W. Wessendorf. 2002. Mu- and delta-opioid receptor mRNAs are expressed in periaqueductal gray neurons projecting to the rostral ventromedial medulla. *Neuroscience* 109:619–634.

Weems, C. F., C. Hayward, J. Killen, and C. B. Taylor. 2002. A longitudinal investigation of anxiety sensitivity in adolescence. *Journal of Abnormal Psychology* 111:471–477.

Weisz, J. R., W. Chaiyasit, B. Weiss, K. L. Eastman, and E. W. Jackson. 1995. A multidimensional study of problem behavior among Thai and American children in school. *Child Development* 66:402–415.

Weisz, J. R., B. Weiss, S. Suwanlert, and W. Chaiyasit. 2003. Syndromal structure of psychopathology in children of Thailand and the United States. *Journal of Consulting and Clinical Psychology* 71:375–385.

Westman, J., J. Hasselstrom, S. E. Johansson, and J. Sundquist. 2003. The influence of place of birth and socio-economic factors on attempted suicide in a defined population of 4.5 million people. *Archives of General Psychiatry* 60:409–414.

Whalen, P. J. 1998. Fear, vigilance, and ambiguity. *Current Directions in Psychological Science* 7:177–187.

Wieser, H. G., and A. M. Siegel. 1993. Relations between EEG of the cortex, thalamus, and periaqueductal gray in patients suffering from epilepsy and pain syndromes. In S. Zschocke and E. J. Speckmann, eds., *Basic Mechanisms of the EEG*. Boston, MA: Birkhauser, pp. 145–182.

Wikan, U. 1989. Illness from fright or soul loss. *Culture, Medicine, and Psychiatry* 13:25–50.

Wild, B., F. A. Rodden, W. Grodd, and W. Ruch. 2003. Neural correlates of laughter and humour. *Brain* 126:2121–2138.

Wilger, E. F. S., and E. B. Kaplan. 1984. The blood and nerve supply to the hand. In M. Spinner, *Kaplan's Functional and Surgical Anatomy of the Hand*. Philadelphia, PA: Lippincott.

Williamson, D. E., K. Coleman, S. Bacanu, B. J. Devlin, J. Rogers, N. D. Ryan, and J. L. Cameron. 2003. Heritability of fearful-anxious endophenotypes in infant rhesus macaques. *Biological Psychiatry* 53:284–291.

Wilson, F. A., and E. T. Rolls. 1993. The effect of stimulus novelty and familiarity on neuronal activity in the amygdala of monkeys performing recognition memory tasks. *Experimental Brain Research* 93:367–382.

Windmann, S., Z. Sakhaut, and M. Kutas. 2002. Electrophysiological evidence reveals affective evaluation deficits early in stimulus processing in patients with panic disorder. *Journal of Abnormal Psychology* 111:357–369.

Wittling, W. 1995. Brain asymmetry in the control of autonomic-physiologic activity. In R. J. Davidson and K. Hugdahl, Eds., *Brain Asymmetry*. Cambridge, MA: MIT Press, pp. 305–357.

Wittling, W., and M. Pfluger. 1990. Neuroendocrine hemisphere asymmetries. *Brain and Cognition* 14:243–265.

Woodward, S. A., M. F. Lenzenweger, J. Kagan, N. C. Snidman, and D. Arcus. 2000. Taxonomic structure of infant reactivity. *Psychological Science* 11:300–305.

Wylie, D. R., and M. A. Goodale. 1988. Left-sided oral asymmetries in spontaneous but not posed smiles. *Neuropsychologia* 26:823–832.

Xiao, E., L. Xia-Zhang, N. R. Vulliemoz, M. Ferin, and S. L. Wardlaw. 2003. Agouti-related protein stimulates the hypothalamic-pituitary-adrenal (HPA) axis and enhances the HPA response to interleukin-1 in the primate. *Endocrinology* 144:1736–1741.

Yamamoto, K., O. L. Davis, S. Dylak, J. Whittaker, C. Marsh, and P. C. van der Westhuizen. 1996. Across six nations: Stressful events in the lives of children. *Child Psychiatry and Human Development* 26:139–150.

Yeh, S. R., R. A. Fricke, and D. H. Edwards. 1996. The effect of social experience on serotonergic modulation of the escape circuit of crayfish. *Science* 271:366–369.

Yilmazer-Hanke, D. M., H. Faber-Zuschretter, R. Linke, and H. Schwegler. 2001. Contribution of amygdalar neurons containing peptides and calcium binding proteins to fear potentiated startle and exploration related anxiety in imbred Roman high and low avoidance rats. *European Journal of Neuroscience* 15:1206–1218.

Zald, D. H., D. L. Mattson, and J. V. Pardo. 2002a. Brain activity in ventromedial prefrontal corte correlates with individual differences in negative affect. *Proceedings of the National Academy of Sciences of the United States of America* 99:2450–2454.

Zald, D. H, M. C. Hagen, and J. V. Pardo. 2002b. Neural correlates of tasting concentrated quinine and sugar solutions. *Journal of Neurophysiology* 87: 1068–1075.

Zech, E., M. M. Bradley, and P. J. Lang. 2002. Affective reactions when talking about emotional events. *Psychophysiology* 39:S90.

Zvolensky, M. J., D. W. McNeil, C. A. Porter, and S. H. Stewart. 2001. Assessment of anxiety sensitivity in young American Indians and Alaskan natives. *Behavior Research and Therapy* 39:477–493.

Difficultness, 41

Disobedience, 16, 30, 63; emotional valence and, 75; geographical variation and, 227; maternal ratings for, 118; socialization threats and, 80

DNA (deoxyribonucleic acid), 50, 227

Dopamine, 65, 66–67, 104, 125, 210–211

Down syndrome, 53

Drugs, 33, 39, 65

Duchenne smiles, 111

Dyslexia, 39

Economic changes, 37–38, 39

Education, 2, 31, 38

Egalitarianism, 2, 36, 38, 224–225

Electric shock experiments, 77–79, 82–83, 84, 85, 87–88; dopamine metabolism and, 125; ethical constraints on, 171–172; heritability and, 230; startle reflex and, 171

Electroencephalogram (EEG), 17, 21, 122, 138–139, 156, 193; beta power and, 205; brain asymmetry and, 123–130; personality types and, 209; skin temperature and, 149; startle reflex and, 177, 181; study findings and, 130–138

Emotions: brain function and, 10; as components of temperament, 5, 40; control of, 30, 103; disguising of, 59; introversion and, 6; in laboratory animals, 210; reaction to the unfamiliar and, 42; somatic components of, 90; valence of, 74–76, 84

Endocrine system, 48

Endogenous wave forms, 158

England, 35, 46, 50, 240

Entorhinal cortex, 74

Environmental influences, 6, 229

Erotic images, reaction to, 49, 173–174

European societies, 36, 37, 225

Event related potentials (ERPs), 157–170, 209

Exogenous wave forms, 158

Extroverts, 6, 32; alpha/beta power and, 123, 138; BAER and, 139; brain hemispheres and, 127; Jung's description of, 2, 218; Q-sorts and, 117

Eyeblink reflex, 172–173, 175, 178, 210

Eye color, 9, 21, 22, 52, 185–188

Eye movements, 129, 163

Facial disfigurement, 60

Facial expression, 11, 19, 54, 84, 121, 128

Families, 3, 29, 38, 114, 222, 240

Fear, 15; amygdalar activation and, 87; bodily sensations and, 102; children rated for fearfulness, 45; cortisol levels and, 68; elusive meaning of, 91–96; eyeblink reflex and, 173; hierarchy of fears, 61; neurochemical reactions and, 49; of novelty, 42, 76–79; philosophical orientation and, 228; serotonin levels and, 65; single or multiple states of, 85–86; startle reflex and, 185; symbolism and, 90

Federmeier, K. P., 195

Felt smiles, 111

Fingertip temperature, 17, 147–149, 154–155; amygdala and, 202; behavior related to, 155; ear temperature and, 156

fMRI. See Functional magnetic resonance imaging (fMRI)

Fox, N. A., 196

Franklin, Rosalind, 50

Freezing, as fear response, 77–78, 88, 89, 91, 93, 171

Freud, Sigmund, 36, 46–48, 79, 98, 240

Frontal lobe, 207

iety disorders and, 97; heritability of IQ scores and, 229; suicide and, 235; uninhibited profile and, 222. *See also* Middle class

Socialization, 5, 24–25, 29, 80, 104, 199

Social phobia, 220–221

Somatic sensations, 101–105

Sorokin, Pitirim, 93–94, 241–242

Sounds, loud, 76, 77, 88, 172, 183

Sperry, Roger, 49

Spiders, fear of, 80, 81, 97

Spinal trigeminal nucleus, 172

Spontaneity, 7, 8, 17, 135, 167; behavioral profile and, 18; induced chronic state of, 194; lack of, 122, 132

Startle reflex, 72, 122, 171–185, 211–213

Strangers, reaction to, 34, 44; brain asymmetry and, 128; great apes and, 188–189; help offered to strangers, 232; observed by strangers, 97, 125; social phobia and, 221

Stress, 60, 61

Striatum, 104, 202

Stroop interference, 212

Suicide, 235

Sulloway, Frank, 120

Symbolism, 90

Sympathetic nervous system, 17, 68, 72, 145–157, 201–202

Tantrums, 30, 54

Task failure, fear of, 100

Temperament, 2, 196–198, 230–234; birth order and, 120–121; as concept, 40–42; as continua or as categories, 51–56; cultural context and, 234–237; cycles in popularity of, 35–40; environmental influences on, 24–27; extreme scores and, 213–216; eye color and, 9, 21, 22; feeling tone and, 217–220; genetics and, 229–230; geography and, 225–229; happiness and, 224–225; life histories and, 3; in literature, 5; measurement of, 50–51; psychopathology and, 220–224. *See also* High-reactive temperament; Low-reactive temperament

Testosterone, 40, 68

Thalamus, 72, 78, 95, 161

Thomas, Alexander, 41, 51, 195

Thorndike, E. L., 5

Tibet, 235–236

Timidity, 15, 27, 44

Traumatic events, 98

Triads procedure, 119–120

Twins, 52, 62, 116

Uncertainty, 33, 111, 157; amygdala and, 125; brain asymmetry and, 125; control over, 28; fear of death and, 93–94; neurochemistry and, 73; parental chastisement and, 25, 30; political catastrophes and, 79; smiles and, 112; unfamiliarity and, 47; varieties of, 91–92; vocational roles and, 221–222

Unfamiliarity, 66–67, 70, 74; animals' reactions to, 77; brain hemispheres and, 124; cultural context and, 236; indifference to, 217; primacy of, 86–91

Uninhibited children, 7–9, 32, 106

United States: adolescents in cultural context, 236; egalitarian premise of, 224; European immigration to, 35–36; happiness defined in, 240; immigration into, 37, 38, 39; individualism in, 36, 37; prevalence of shyness in, 25; psychologists in, 36–37

Vagal tone, 56, 102, 202, 205; cardiovascular system and, 147, 149–152; of "strong silent types," 191

Valence, emotional, 74–76